NEUROSCIENCE RESEARCH PROGRESS

ADVANCES IN NEUROPATHY RESEARCH

MICHAEL O. BROWN
AND
BARBARA A. BUDD
EDITORS

NOVA BIOMEDICAL

New York

Library of Congress Cataloging-in-Publication Data

ISBN: 978-1-62618-400-8

Library of Congress Control Number: 2013934501

Published by Nova Science Publishers, Inc. † New York

Contents

Preface

Neuropathies are a collection of disorders that occur due to damage, destruction or dysfunction to the nerves of the central or peripheral nervous system. The causes may be traumatic injury, infections, metabolic disorders or chemical insults to the nervous tissue resulting in crippling complex disease conditions. In this book, the authors discuss the current understanding of the status of neuropathy research and the therapeutic role of phytochemicals and spices; cardiac autonomic dysfunction in Type 2 Diabetes Melliuts; objective investigation of multiple sclerosis related optic neuritis; neuropathy induced by organophosphorus compounds; and the role of neuromuscular ultrasound in the diagnostics of immune-mediated neuropathies.

Chapter 1 - Neuropathy is a collection of disorders that occurs due to the damage to the nerves. Although more commonly referred to the damage to the peripheral nervous tissue, neuropathy can be damage, destruction or dysfunction of either the central or peripheral nervous system. The causes may be traumatic injuries, infections, metabolic disorder and chemical insults to the nervous tissue resulting in a crippling complex disease condition. The symptoms of neuropathy are broadly categorised as sensory and motor deficits, comprising of numbness, loss of sensation, hypo or hyper sensation and co-ordination and locomotor dysfunctions. The cardinal feature of neuropathy is pain (often referred to as neuropathic pain) is mostly described as burning, shooting, stabbing and cruciate. Neuropathies and the related pain are often classified according to the underlying etiology, eg. diabetic neuropathy, post-herpetic neuralgia, trigeminal neuralgia, cancer related neuropathy, HIV neuropathy, central poststroke pain, phantom limb pain and spinal cord injury. The pathophysiological mechanisms of neuropathic diseases comprise various biochemical alterations in brain, spinal cord and/ or peripheral nerves. Current understanding of the pathology encompasses oxidative stress, inflammatory reactions, mitochondrial dysfunction and apoptosis in the nervous tissue. Several well accepted animal models have been developed to understand the pathophysiology of neuropathy. Both *in vitro* and *in vivo* models have been deployed as tools to understand the network of mechanisms that underlie the pathophysiology of neuropathy. While several primary cell cultures (eg. primary cortical cell culture, DRG neurons, CGC culture) and cell lines (eg. PC 12, N27, SH-SY5Y) are used as *in vitro* models, rodent models are extensively employed to obtain *in vivo* data. Although, neuropathy is idiopathic, diabetic neuropathy is the most common and 60-70% of people with diabetes exhibit some form of neuropathic signs. While many therapeutic approach are being employed to manage neuropathy in general and neuropathic pain in particular, many pharmacological treatments pose a concern for their effectiveness and side effects. In recent times, a body of evidence suggests promising beneficial effects employing spices and phytochemicals to alleviate neuropathic as well specific neurodegenerative conditions. Researchers world-wide are constantly exploring newer approaches and attempts employing various bio-actives such as curcumin, eugenol, etc. (either alone or in combination) in various models of neuropathy have yielded interesting data owing to their specific effects on multiple targets. This review will discuss the current evidence related to the possible therapeutic usage of spices and phytochemicals in the management of neuropathy in humans.

Chapter 2 - Diabetes mellitus is a metabolic disorder affecting various organs of the body. In many patients with diabetes mellitus, typical symptoms associated with the disease manifest only after sufficient cumulative effect of diabetes mellitus. As a result, diabetes-related complications and associated clinical conditions may be present by the time it is clinically diagnosed. One such diabetic complication which remains subclinical is autonomic neuropathy. Diabetic neuropathy is associated with morbidity and mortality. In the light of increased mortality rate associated with diabetic autonomic neuropathy, and its possible involvement in other complications it is of utmost importance to assess cardiac autonomic function periodically for long-term treatment strategies. Among the different non-invasive techniques for assessing autonomic status, heart rate variability has emerged as a major tool in the diagnostic and monitoring armamentarium.

Currently, based on the recommendations of task force of the European Society of Cardiology and the North American Society of Pacing and Electrophysiology time domain measures of heart rate variability are typically obtained from a routine 24-h ambulatory electro-cardiogram. However heart rate variability parameters derived from much lesser duration may be more appropriate for routine clinical assessment of cardiac autonomic dysfunction. This article portrays insights from one minute electrocardiogram studies on assessment of cardiac autonomic dysfunction in type 2 diabetes mellitus; wherein, heart rate response and Inspiratory: Expiratory ratio are derived from lead II electrocardiogram while deep breathing at 6 respiratory cycles per minute; rMSSD and SDNN are derived from lead II electrocardiogram while breathing spontaneously in resting supine position.

Analysis of heart rate response to deep breathing at 6 respiratory cycles per minute at baseline reveals that, among the clinical characteristics of type 2 diabetes mellitus, somatic neuropathy is an independent predictor of cardiac parasympathetic dysfunction. Age and female gender are the other predictors of reduced heart rate response to deep breathing in type 2 diabetes mellitus.

Monitoring cardiac autonomic function over one year period reveals that cardiac autonomic function deteriorates with time and essentially in presence of somatic neuropathy and cardiovascular disease. rMSSD an indicator of parasympathetic activity detects deterioration in cardiac autonomic dysfunction coexisting with somatic neuropathy and cardiovascular disease. But Expiratory: Inspiratory ratio another indicator of parasympathetic function is a suitable marker in monitoring cardiac autonomic dysfunction coexisting with somatic neuropathy. Similarly, SDNN an indicator of parasympathetic and sympathetic selectively quantifies cardiac autonomic dysfunction coexisting mainly with cardiovascular disease. Thus these findings suggest the possible utility of time domain measures of heart rate variability derived from one minute electrocardiogram in routine clinical assessment of cardiac parasympathetic dysfunction in type 2 diabetes mellitus.

Chapter 3 - Optic neuritis (ON) is an inflammatory, infective or demyelinating process, affecting the optic nerve. Retrobulbar neuritis, in which the optic disc appearance is normal, at least initially, is the most frequent type in adults and is often associated with multiple sclerosis. Multiple sclerosis is a complex neurodegenerative disorder which can cause visual dysfunction in 80% of patients during the course of the disease. Axonal loss is considered to be responsible for visual disability in ON associated with MS rather than demyelination with preservation of axons. The value of magnetic resonance imaging (MRI) in the diagnosis of multiple sclerosis is well established, as MRI provides information about disease burden, emphasizing on inflammation and demyelination. However, there is a poor correlation between quantification of axonal and neuronal loss by MRI and functional disabilities evaluated by clinical measures.

Optical coherence tomography (OCT) is a non-invasive high-resolution method, routinely used in ophthalmology to visualize and quantify the layers of the retina. It provides information on optic nerve head topography, retinal nerve fiber layer (RNFL) thickness and macular volume, which correlate with axonal loss. Several studies found a significant reduction in RNFL thickness in eyes affected by multiple sclerosis with or without optic neuritis, compared to control eyes. RNFL thickness was correlated with visual acuity, visual field, contrast sensitivity and colour vision.

In addition to OCT, which provides quantitative measurements of the RNFL thickness, visual evoked potentials (VEP) offer a functional assessment of the integrity of the optic nerve in ON. OCT findings were correlated with amplitude of visual evoked potentials (VEP), but not with latency. Nowadays the new multifocal VEP (mVEP) allows the detection of multiple individual VEP signals across the visual field up to eccentricity of 20° to 30°. As the changes of optic nerve are more diffuse than originally thought based on standard perimetry, mVEP was found to be helpful in predicting progression to future multiple sclerosis.

Chapter 4 - Organophosphorus-induced delayed neuropathy (OPIDN) was observed in more than 70,000 human cases in the last 75 years. This type of neuropathology has been known since 1930s when the tuberculosis patients were treated with phosphocreosote, which is an undefined mixture of esters derived from phosphoric acid and phenols. In the 1930s, thousands of cases of paralysis occurred in the United States due to the consumption of Jamaica ginger extract contaminated with tri-ortho-cresyl phosphate (TOCP), which is the compound used as inducer in experimental studies of delayed neuropathy induced by organophosphates (OPs). OPIDN is characterized by a distal axonopathy, central and peripheral Wallerian-type which develops from 8 to 14 days after intoxication by a neuropathic OP. Among the OPs that cause OPIDN include phosphates, phosphonates and phosphoramidates. TOCP, methamidophos, mipafox, dichlorvos, trichlorfon and leptophos are examples of some compounds that have been reported as inducers of OPIDN. The methamidophos (O,S-dimethyl phosphoramidothioate) has been reported the pesticide as more often causes OPIDN in humans. To develop this neuropathy is required at least 70-80% of inhibition of an esterase known as neuropathy target esterase (NTE). NTE was originally identified by M. K. Johnson as target of neuropathic OPS. NTE is a membrane-bound enzyme with 1327 amino acids belonging to the group of esterases with a molecular weight of

155 kDa and its physiological role is not yet established. The usual symptoms of OPIDN are early cramps, muscle pain, tingling, numbness, weakness and progressive distal burning more present in feet than in hands. Then, a progressive weakness in the legs is evident. More severe cases can lead to paralysis, respiratory problems and death. Several animal species are sensitive to OPIDN. In this list include chickens, cats, dogs, horses, sheep, pigs, monkeys and man. The hens are the animal model for studying this neuropathy, due to their sensitivity and clinical signs were very similar to those seen in humans. Furthermore, the progress of these signals is easily observed and correlated with pathological changes.

Chapter 5 - The immune-neuropathies are a diverse group of syndromes. Their diagnosis and classification is based in classical cases, on the distribution pattern of peripheral nerve impairment and the results of nerve conduction studies. In some cases though, the clinical presentation may be subtle, so that extended differential diagnostic workup is needed, including cerebrospinal fluid examination, laboratory tests and in some cases nerve biopsy or magnetic resonance imaging of proximal nerves or roots.

Electrophysiological testing remains nowadays fundamental not only for the diagnosis, but also for the follow-up and measurement of response to immune-treatment in these type of neuropathies. New challenges arise though, on how best to acquire a static and dynamic imaging of the relevant nerve structures, aiming to provide a complementary and holistic approach to the nerve impairment. As ultrasound technology and resolution continoue to improve, nerve imaging has become more detailed and allows visualisation of nerve fascicles and small structural changes, which will move nerve imaging closer to the type of detail that can be seen in biopsy material. Neuromuscular ultrasound has been able to detect thickened or swollen roots, peripheral nerves or plexus, findings that are consistent with ongoing inflammation, especially in cases of chronic inflammatory demyelinating polyneuropathy (CIDP). Similar findings have been described also in other immunomediated neuropathies such as *Guillain–Barré* syndrome (GBS), multifocal motor neuropathy (MMN) and multifocal acquired demyelinating sensory and motor neuropathy (MADSAM). Although this figures are rewarding to both physicians and patients, it remains unknown, whether ultrasound could be a reliable method for follow up and measurement of response to immune-treatment.

This review provides a timely update on the ultrasound findings of different immune-neuropathies, while possible future possibilities of neuromuscular ultrasound are also discussed.

In: Advances in Neuropathy Research ISBN: 978-1-62618-400-8
Editors: M. O. Brown and B. A. Budd

Chapter 1

Current Understanding on the Status of Neuropathy Research and the Therapeutic Role of Spices and Phytochemicals: An Overview

*Sathya N. Prasad and Muralidhara**
Department of Biochemistry and Nutrition
CSIR-Central Food Technological Research Institute(CFTRI)
Mysore, India

Abstract

Neuropathy is a collection of disorders that occurs due to the damage to the nerves. Although more commonly referred to the damage to the peripheral nervous tissue, neuropathy can be damage, destruction or dysfunction of either the central or peripheral nervous system. The causes may be traumatic injuries, infections, metabolic disorder and chemical insults to the nervous tissue resulting in a crippling complex disease condition. The symptoms of neuropathy are broadly categorised as

* Corresponding Author: Dr Muralidhara, Chief Scientist, Dept. of Biochemistry and Nutrition, CSIR- Central Food Technological Research Institute, Mysore - 570020, India. TEL: 91-821-2514876; FAX: 91-821-2517233, E-mail address: mura16@yahoo.com.

sensory and motor deficits, comprising of numbness, loss of sensation, hypo or hyper sensation and co-ordination and locomotor dysfunctions. The cardinal feature of neuropathy is pain (often referred to as neuropathic pain) is mostly described as burning, shooting, stabbing and cruciate. Neuropathies and the related pain are often classified according to the underlying etiology, eg. diabetic neuropathy, post-herpetic neuralgia, trigeminal neuralgia, cancer related neuropathy, HIV neuropathy, central poststroke pain, phantom limb pain and spinal cord injury. The pathophysiological mechanisms of neuropathic diseases comprise various biochemical alterations in brain, spinal cord and/ or peripheral nerves. Current understanding of the pathology encompasses oxidative stress, inflammatory reactions, mitochondrial dysfunction and apoptosis in the nervous tissue. Several well accepted animal models have been developed to understand the pathophysiology of neuropathy. Both *in vitro* and *in vivo* models have been deployed as tools to understand the network of mechanisms that underlie the pathophysiology of neuropathy. While several primary cell cultures (eg. primary cortical cell culture, DRG neurons, CGC culture) and cell lines (eg. PC 12, N27, SH-SY5Y) are used as *in vitro* models, rodent models are extensively employed to obtain *in vivo* data. Although, neuropathy is idiopathic, diabetic neuropathy is the most common and 60-70% of people with diabetes exhibit some form of neuropathic signs. While many therapeutic approach are being employed to manage neuropathy in general and neuropathic pain in particular, many pharmacological treatments pose a concern for their effectiveness and side effects. In recent times, a body of evidence suggests promising beneficial effects employing spices and phytochemicals to alleviate neuropathic as well specific neurodegenerative conditions. Researchers world-wide are constantly exploring newer approaches and attempts employing various bio-actives such as curcumin, eugenol, etc. (either alone or in combination) in various models of neuropathy have yielded interesting data owing to their specific effects on multiple targets. This review will discuss the current evidence related to the possible therapeutic usage of spices and phytochemicals in the management of neuropathy in humans.

1. Introduction

Any damage to the nervous system and the associated dysfunction is collectively called neuropathy. It is a collection of disorders related to the nervous system with different causal agents. It can be a disease of the peripheral nervosu system (PNS) and/ or central nervous system (CNS) correspondingly referred to as peripheral neuropathy and encephalopathy.

Under neuropathic condition, voluntary or involuntary functions or both may be affected however the word neuropathy is more often referred to peripheral neuropathy.

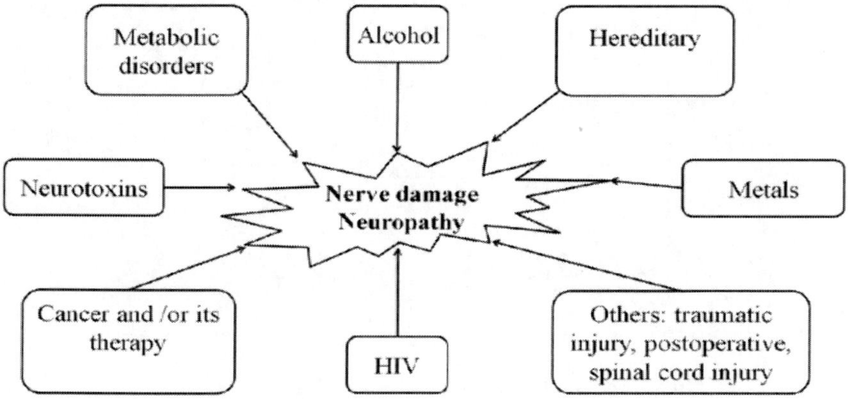

Figure 1. A scheme representing a wide spectrum of possible causes for neuropathy.

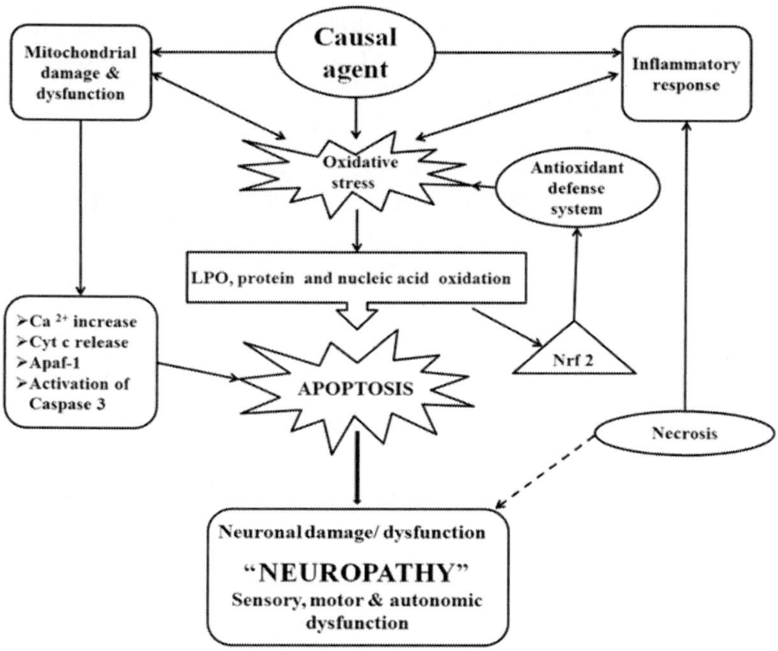

Figure 2. Possible mechanisms which are involved in the development of neuropathy.

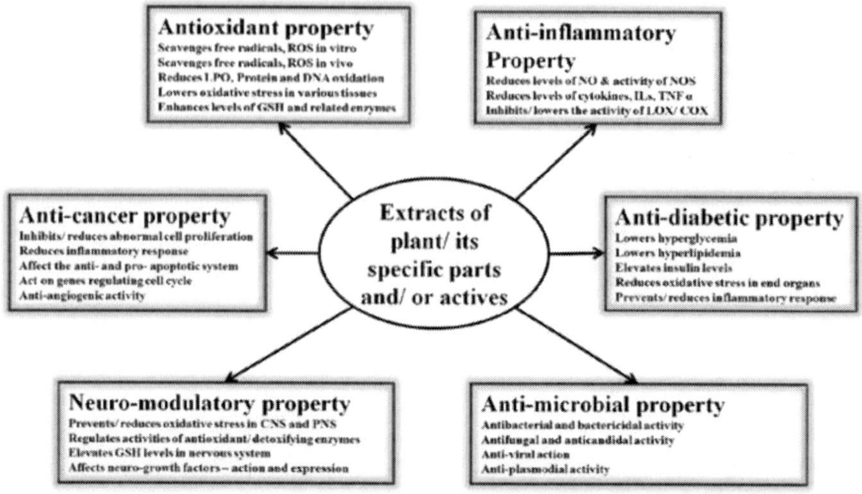

Figure 3. Plethora of various biological activities which may significantly to contribute towards the therapeutic property of herbal extracts and their actives.

Neuropathic pain generally associated with peripheral nerve injury is characterized by the sensory abnormalities such as unpleasant and abnormal sensation (dysesthesia), an increased response to painful stimuli (hyperalgesia) and pain in response to a stimulus that does not normally provoke pain (allodynia) (Wolf and Mannion, 1999). The intensity of neuropathy varies with the causative agent and is assessed in many cases based on the degree of symptoms experienced by the subject. Peripheral neuropathic pain is frequently observed among patients with diabetic neuropathy, cancer, acquired immune deficiency syndrome (AIDS), leprosy, post-herpetic neuralgia, central poststroke pain and traumatic injuries (Garg and Adams, 2012; Mannelli et al., 2009) (Figure 1).

The pathophysiology associated with different types of neuropathies studied so far deals with the biochemical alterations and morphological aberrations in brain, spinal cord and peripheral nerves. Some of the cellular changes under neuropathic condition include oxidative stress, mitochondrial dysfunction, and alteration in synthesis and / or transport of neurofilaments, altered levels of neurotrophic factors and activation of apoptotic cascade (Figure 2).

These factors result in motor disorders, altered sensibility and the most importantly in neuropathic pain. The vital physiological factors are discussed briefly below.

1.1. Oxidative Stress and Inflammatory Response

Oxygen supports survival and function of the cell, however, hyperoxia induces toxicity including neurotoxic effects (Chavko et al., 2003). The nervous tissue (brain, spinal cord and the peripheral nerves) is very active metabolically and hence requires uninterrupted oxygen. It utilizes about 20% of the total oxygen taken up at rest and more while active state (Halliwell 2006). The main reason for higher uptake is the higher requirement of energy/ ATP to maintain neuronal intracellular ion homeostasis as propagation of action potentials and neuro-secretion is driven on a continuous basis. Depriving the brain of its rich oxygen supply even for short intervals of time often has severe and irreversible damage -ischemia (Reiter et al., 2010). Such morphological and physiological consequences occurring in both neurons and glia can be fatal. While oxygen supply is indispensible unit of brain survival/ function, it indirectly contributes to its destruction and death over time. This is because there is always a basal level of oxidative damage to all the vital biomolecules (DNA, lipids and proteins) as a small percentage of the oxygen is metabolized to derivatives that gradually erode and destroy essential molecules. This eventually accumulates and hampers normal functioning of the cell (Reiter et al., 2010; Kirkwood, 2005). In the process, highly reactive and unstable entities called the reactive oxygen species (ROS) and reactive nitrogen species (RNS) are produced. Some of the examples include superoxide radical, hydrogen peroxide (H_2O_2), hydroxyl radical, nitric oxide and nitrate. Some of these markers also act as triggers to the immune response leading to inflammatory reactions as a defence response against infection (Halliwell and Gutteridge 2006). When such reactive species go unchecked/ accumulate as in case of an allergic response, apoptosis is triggered and cells eventually die. Hence it is necessary that cells must be equipped with means to quench the effect of such reactive species so that their damaging effect on cell constituents and function is prevented. With the fact that most neurons cannot divide or be replaced, the tissue function is hampered as a whole following nervous dysfunction. Hence to destruct/ scavenge the reactive species or convert them to metabolically non-reactive and /or more soluble form, a set of reaction occurs in every cell, this is collectively called as antioxidant defense system. One of the mechanism by which cells combat oxidative insult is through increased transcription of genes containing the antioxidant response element (ARE). ARE in turn regulates many cytoprotective genes *via* the transcription factor Nrf2 (Calkins et al., 2008). The antioxidant defence system comprises of various antioxidant molecules such as reduced glutathione

(GSH), thioredoxin and vitamin C, and enzymes such as superoxide dismutase, catalase, glutathione-S-transferase and glutathione peroxidase (Halliwell 2006; Maiese et al., 2010).

Mammalian brain is known to be more sensitive to such oxidative and inflammatory insults as it is not effectively equipped with antioxidant defence. One of the primary reasons as mentioned earlier is the higher degree of oxygen consumption by brain (Reiter et al., 2010; Halliwell 2006). Secondly, although vitamin C is a powerful antioxidant, its presence in large amounts in different brain regions makes it a pro-oxidant with the availability of free iron (Reiter et al., 2010; Burdo and Connor, 2003). Thirdly, brain metabolism results in the production of considerably higher levels of H_2O_2. Further the cytochrome P450 present in different brain regions also generates higher amount of ROS. Fourthly, the neuronal membrane is rich in poly unsaturated fatty acids (PUFAs) and is prone to oxidative damage by the reactive species leading to the loss of membrane integrity (Reiter et al., 2010; Halliwell 2006). Further, the products of lipid peroxidation such as hydroxynonenal are cytotoxic in nature (Ong et al., 2000). Loss of membrane integrity can also lead to disturbances in the ionic homeostasis, especially the influx of Calcium (Ca^{2+}) leading to increased levels of Ca^{2+}. This can bring about (i) activation of neuronal nitric oxide synthase (nNOS), increasing the levels of NO (ii) activation of phospholipase A2, further damage to the organelle and plasma membrane (iii) activation of calpains, a family of Ca^{2+} stimulated proteinases that attack cytoskeleton and neurofilaments (iv) inactivation of Glutamate transporters (Halliwell 2006; Fonfria et al., 2005). Such events invite inflammatory response either *via* NO or 5-lipoxygenase pathway or cyclooxygenase pathway.

1.2. Mitochondrial Dysfunction and Apoptosis

Mitochondria (Mt) utilize more that 80% of the total oxygen taken up by the human body that is the reaction catalysed by cytochrome oxidase that results in the formation of water (Halliwell, 2006). Oxidative stress affects mitochondrial function, either by damaging the membrane followed by an imbalance in the ionic homeostasis or allowing the entry of neurotoxins. Oxidative stress may also affect the continuous supply of oxygen or substrates of the electron transport chain (ETC), further damaging the whole cell function.

The neuronal and astroglial cells can display a differential susceptibility to oxidant exposure. While astrocytes can up regulate GSH availability in wake of mitochondrial damage, neuronal cells manage by uptake of GSH precursors from the astrocytes (Bolanos et al., 1995). However, when the redox balance shifts to oxidative stress, Mt is damaged. The redox carriers of the respiratory chain to the final electron acceptor – oxygen, get affected. Neuronal Mt significantly contributes to the ROS levels by generating superoxide molecules *via* the ETC (Halliwell, 2006). Normally 0.1% of the total oxygen consumption leaks from the ETC to generate ROS, however, uncoupling of the respiratory chain, oxidative damage of the ETC members and the phospholipid (eg. cardiolipin) of the membrane will add more oxidative burden to the cell (Pope et al., 2008; Rolo and Palmiera, 2006).

Each mitochondrion contains 2-10 copies of circular, supercoiled, double stranded DNA found unprotected within the inner mitochondrial membrane. The close proximity of Mt DNA to ROS generating sites makes it more vulnerable to oxidative damage, in turn affecting the proteins expressed by them (Rolo and Palmiera, 2006).

Approximately 90% of cellular ROS arises from the Mt and are mainly from the oxidative phosphorylation. Detoxification of highly reactive super oxide anions by SOD, an antioxidant enzyme present in the mitochondrial matrix, results in hydrogen peroxide production, which in turn is reduced to water, catalysed either by glutathione peroxidase or catalase (Lee et al., 2012). Oxidative damage, either of oxidative phosphorylation/ ETC proteins or of long chain PUFA in mitochondrial membranes results in loss of the mitochondrial membrane potential (MMP). As a result, the mitochondrial pathway of apoptosis is initiated by the release of cytochrome c and apoptosis inducing factor (AIF) into the cytosol accompanied with disrupted Ca^{2+} homeostasis. This in turn activates caspases and down-stream pathway of apoptosis (Green and Reed, 1998).

Anti-apoptotic Bcl-2 protein is mainly localized in the outer mitochondrial membrane and upon release it stabilizes the inner mitochondrial membrane against several injuries (Reed, 1998). Decreased Bcl-2 expression and increased expression of pro-apoptotic Bax have been associated with activation of c-Jun N-terminal kinase (JNK) (Eckert et al., 2003; Cory et al., 2003). The resulting mitochondrial dysfunction, especially in from of severe changes in MMP, is believed to be crucial for the onset and progression of neurodegenerative diseases.

1.3. Neurofilaments

The neuronal cytoskeleton is composed of three interconnected structures Viz., the actin microfilaments (MFs), microtubules (MTs) and intermediate filaments (IFs). IF forms a large family of proteins of various sizes (40–280 kDa) and primary structures. Neurofilaments (NF) are the major IFs present in adult neurons and their expression is restricted to neuronal cell types (Perrot et al., 2008). Neurons express differentially several IF proteins depending on the developmental stage or based on their localisation in the nervous system. IFs expressed by mammalian neurons consist of α-internexin, the NF triplet proteins (NFL, NFM and NFH for low-, medium- and high-molecular-weight subunits respectively), peripherin, synemin and nestin. α-internexin and peripherin are co-assembled with NF. They increase the axonal calibre of myelinated axons and consequently their conduction velocity contributing to the dynamic properties of the axonal cytoskeleton during neuronal differentiation, axon outgrowth, regeneration and guidance (Nixon and Shea, 1992). Phosphorylation of NFH slows neurofilament transport, and this is due to increased pausing in neurofilament movement (Ackerley et al., 2003). The frequent disorganisation of NF network observed in several neuropathologies support their contribution to neuronal integrity. However, despite the presence of NF mutations found in some patients, the exact relations between these mutations, the abnormal NF organisation and the pathological process remains unclear. Perturbations of their metabolism and organisation are frequently associated with various neurodegenerative diseases, including amyotrophic lateral sclerosis (ALS), Alzheimer's disease (AD) and Charcot-Marie-Tooth disease (CMT).

1.4. Neurotrophins and Neuregulins

Neurotrophins or neurotrophic factors are a family of endogenous substance that regulate the growth and survival of the CNS and PNS. Basically they support the growth/ maintenance of the neurons as well as growth of neurites for their complete network and thus are indispensible unit for the development/functioning of the nervous system. The family of neurotrophins include nerve growth factor (NGF), brain derived neurotrophic factor (BDNF), neurotrophin-3 (NT-3), neurotrophin -4/5 (NT-4/5) and neurotrophin-6 (NT-6). NGF, ubiquitously present in the nervous system, promotes survival and neuritogenesis in primary neurons of the CNS and PNS (Sofroniew et al.,

2001). Interruption of NGF and BDNF signalling results in the apoptosis and neuronal death (Matrone et al., 2008). Likewise, both NGF and BDNF render protection against neurotoxin -induced apoptosis in neural stem cells (Nguyen et al., 2009). Insulin like growth factors (IGF-I, IGF-II), cytokine like growth factors (CTNF) and vascular endothelial growth factor (VEGF) are also considered to be neurotrophic factors. Alteration in the synthesis and expression levels of some of these growth factors in the peripheral nervous system is observed in nervous disorders.

The neuregulins are a family of growth and differentiation factors with a wide range of functions in the nervous system (and heart) (Raabe et al., 2004). A bidirectional mechanism regulates the release of neuregulins in response to neurotrophic factors produced by both neuronal targets and supporting glial cells. Both neuregulins and nerotrophic factors produce a balanced intercellular signalling system that contributes to the normal development as well as maintenance of the mature nervous system (Falls, 2003). Neuregulin-1 (NRG1) is one of the extensively studied neuregulins, known to participate at all levels of neuronal function such as peripheral nerves, neuromuscular junction, the spinal cord, brainstem and neocortex. Recent evidence suggests that neuregulin signalling plays important roles in many neurological disorders including multiple sclerosis, traumatic brain and spinal cord injury, peripheral neuropathy and schizophrenia (Esper et al., 2006).

2. Types of Neuropathy

2.1. Diabetic Neuropathy (DN)

Both type I and type II diabetes have largely irreversible and devastating effect on small and large blood vessels. One of the most common complications of diabetes associated macrovascular diseases is DN. The WHO estimates that 366 million people world-wide will suffer from diabetes by the year 2030 (Wild et al., 2004). More than 60% of diabetic people (that is an estimated population of 219 million by 2030) develop/ exhibit some form of neuropathy resulting in sensory loss, pain and autonomic dysfunction (Little et al., 2007; Brownlee, 2004; Wild et al., 2004). DNs are a family of nerve disorders caused by persistent hyperglycemia. Diabetics can over progression of time develop nerve damage throughout the body. DN affects both somatic and autonomic divisions of the PNS and a growing body of evidence reveals that the spinal cord and higher CNS may also be damaged. Since diabetes not only damages peripheral neurons, but also impairs their capacity to regenerate, DN is also considered as a form of neuron degeneration (Zochodne et al., 2008).

DNs are further classified based on diagnosis and/ or extent of associated symptoms (Little et al., 2007; Fernyhough et al., 2010). Initially, these include numbness and tingling in toes or feet and the symptoms spread to knee followed by fingertips with longer duration of diabetic condition. Other symptoms include indigestion, nausea, vomiting; diarrhea or constipation; urination; sexual dysfunction (erectile dysfunction – men, vaginal dryness – women); dizziness or faintness (due to drop in blood pressure and/ or alteration in blood glucose levels); wasting of muscles and weakness. Diabetes associated neuropathy is sometimes also referred to as polyneuropathy based on the nature of symptoms. Most of these symptoms such as numbness, tingling and progressive balance difficulties are due to loss of thickly myelinated neve fibres (large fibre symptoms) and spontaneous shooting/ stabbing pains, hot or cold burning sensations and allodynia are due to dysfunction of the thinly myelinated and unmyelinated nerve fibres (small nerve fibre sensations) (Little et al., 2007; Fernyhough et al., 2010).

Various mechanisms are known to be involved age in DN since hyperglycemia can activate many signalling mechanisms in cells. These can be explained as follows : Excess glucose (i) shunts to the polyol pathway that depletes cytosolic NADPH and subsequently GSH (ii) undergoes autooxidation to produce advanced glycation end products (AGEs) that impair protein function and also activate receptors of AGEs that use ROS as second messengers (iii) activates protein kinase C (PKC) and its down-stream signalling, that further increases hyperglycemia and also exacerbates tissue hypoxia (iv) overload and slowing of electron transfer chain (ETC) lead to escape of reactive intermediates to produce superoxides as well as activation of NADH oxidase that also produces superoxide (Vincent et al., 2004a; Schemmel et al., 2010). These result in further enhancement of oxidative stress, inflammatory reactions, mitochondrial dysfunction, impaired energy metabolism, discrepancy in calcium ion homeostasis and apoptosis (Vincent and Feldman, 2008; Fernyhough et al., 2010; Kamboj and Sandhir, 2011). Each of these leads to oxidative damage to vital biomolecules such as proteins, lipids and nucleic acid (with elevated levels of different types of ROS). This impairs their diverse functional and structural role as well as transcription and translation function of the genes (Rolo and Palmeira, 2006; Obrosova 2009).

2.2. Cancer and Chemotherapy

Cancer related neuropathic pain syndromes/ peripheral neuropathy are common and serious complications of a patient's primary health and treatment (Cleeland et al., 2010; Gilardini et al., 2012). Cancer related neuropathy whether a result of nerve damage by the tumour itself and/ or by treatment, causes significant morbidity. Tumour can have a direct impact on the nerves, such as compression of the nerves in the vicinity of tumour growth, through mechanical distortion, loss of blood supply leading to ischemia that may be followed by an immune response creating an inflammatory microenvironment (Bennet, 2010). Treatment of any type such as surgery, radiation or chemotherapy does result in peripheral neuropathy (Bennet, 2010; Mannelli et al., 2012). Studies have shown that chemotherapy -induced peripheral neuropathy (CIPN) is a dose-limiting side effect of anticancer therapy which significantly affects the quality of patients' life. At times, CIPN can produce high intensities of pain, such that continuing treatment for cancer becomes difficult. Despite the encouraging findings about the anticancer mechanisms of action of cisplatin, paclitaxel, vincristine, bortezomib and others to a great extent; their neurotoxic effects are not yet be deciphered completely (Gilardini et al., 2012). The symptoms of neuropathy among cancer patients include acute to severe sensory and motor dysfunction that include signs of neuropathic distress such as paresthesia, dysesthesia, burning and pain. However, the degree of these signs differs from patients to patients also with the type of cancer and the type of anticancer agents administered to them.

Often it is difficult to distinguish neuropathic pain from nociceptive though the underlying pathophysiology is different. Pain and other symptoms of cancer related peripheral neuropathy are not clearly understood. While the reversal of such neuropathy may be slow and is possible once the course of administration of the anticancer drug is completed/ withdrawn; for eg. thalidomide (Kannarket et al., 2007), but in case of certain drugs such as cisplatin, withdrawal does not help in improving the neuropathic status and the damage persists (Albers et al., 2011; Windebank et al., 1999). Cisplatin is toxic to the nervous system and exhibits preferential uptake by the dorsal root ganglion and a produces a dose related large nerve fibre sensory neuropathy (Albers et al., 2011). The pathophysiology of CIPN includes induction of mitochondrial dysfunction in sensory neurons, oxidative stress, inflammatory reactions, morphologically identified axonal and neuronal aberrations (Xiao and Bennett, 2012; Mannelli et al., 2012; Zheng et al., 2011a; Albers et al., 2011; Abad et al., 2011; Carozzi, 2010). Platinum based anticancer agents such as cisplatin, have been found to inhibit axonal growth, induce abnormalities in the neucleoli of spinal root ganglion and sensory ganglion neurons/ cell body accompanied with alterations in peripheral nerve - sciatic nerve (Albers et al., 2011; Carozzi, 2010). Degeneration of peripheral nerve fibres (A and C) and peripheral nerve axon occurs in response to the dosage and duration of the chemotherapeutic agent (Zheng et al., 2012). Significant abnormalities in both sensory and motor axons were evident in the nerve excitability studies of oxaliplatin treated patients (Krishnan et al., 2005, 2006). In addition, autonomic nerves may also be damaged affecting some of the involuntary actions such as cardiovascular and gastro-intestinal function. Although this is difficult to diagnose, these aberrations are together called as autonomic neuropathy.

2.3. AIDS – HIV Associated Neuropathy

Many studies have reported that human immunodeficiency virus (HIV) infection, causal agent of acquired immune deficiency syndrome has a pathophysiological implication on the nervous system (Ferrari et al., 2006). The neurological complication resulting from HIV infection is referred to as HIV – associated distal sensory neuropathy (HIV-SN) (Irish et al., 2009; Phillips et al., 2010). It is a progressive disorder that develops during late HIV infection and is rare in otherwise healthy seropositive patients. HIV – 1, one of the retroviruses, causes neurodegenerative diseases such as HIV associated dementia (HAD) or NeuroAIDS. The disease processes that result in neurodegeneration is at least in part due to (i) a dys-regulation of inflammatory cytokines (tumor necrosis factor, interleukins – IL-1, IL-6) (ii) recruitment of or local infiltration of immune cells (activated monocyte-macrophages) to the CNS (iii) distal epidermal denervation and (iv) functional properties of viral proteins such as gp120 (envelope protein), gp41, Tat, VPr mediated neuronal damage and subsequent motor and neurocognitive impairment (Irish et al., 2009; Smith, 2011). Several evidence suggest that gp120 disrupts calcium homeostasis, induces disruption of mitochondrial membrane integrity which in turn results in release of cytochrome c that activates caspases and endonucleases ultimately bringing about neuronal apoptosis (Lannuzel et al., 1995; Mattson et al., 2005). This retroviral-induced neurological disease, leads to devastating and progressive neurodegenerative consequences. This result in HIV-SN, affecting about 90% of HIV infected individuals who are experiencing potentially debilitating neuropathic pain (Phillips et al., 2010).

Like DN, patients may be asymptomatic or experience allodynia in feet, distal numbness and weakness, accompanied with autonomic dyfunction at a later stage with urogenital and intestinal functions being affected. Some of the histological findings include axonal degeneration, demyelination of nerve fibres, peri-vascular infiltration by T lymphocytes and macrophages; and mild loss of dorsal root ganglion neurons. Unfortunately, HIV-SN prevalence has increased (or despite of) the introduction of otherwise successful anti-retroviral therapy. Although, anti-retroviral therapy has had a great impact on life expectancy of patients with reduction in their viral load and peripheral disease, it has increased the prevalence of HIV-SN (Childs, et al., 1999; Symth et al., 2007).

2.4. Neurotoxins

Neurotoxins are compounds that can damage the nervous system either specifically the CNS and/ or the peripheral nerves. These include metals as well as naturally occurring and synthetic compounds. Common examples include cyanide, rotenone, acrylamide, ethanol, toluene, 6-hydroxy dopamine, aminolevulinic acid, cyanide, kainic acid, metals (eg., lead, iron, aluminium and mercury), methamphetamine and polychlorinated biphenyls.

2.4.1. Acrylamide (ACR)

ACR, a vinyl monomer, is a human neurotoxin while its polymerised form is innocuous. ACR has wide industrial applications such as waste water management, ore processing, dye synthesis and cosmetics and is used in gel electrophoresis in biochemistry and molecular biology laboratories (LoPachin, 2004; Yener and Kalipci, 2009). Carbohydrate foods processed at high temperatures, such as potato chips, French fries, cookies and biscuits contribute to the dietary intake of ACR (Stadler et al., 2002; Tareke et al., 2002). ACR neurotoxicity (among both laboratory animals/humans) is characterized by ataxia and distal skeletal muscle weakness (Yu et al., 2006). A recent study estimated a Tolerable Daily Intake (TDI) 40 μg/kg bw/d for ACR induced neurotoxicity (Tardiff et al., 2010). However, ACR exposure in industrial settings is estimated to be much higher. ACR intoxication in rodent models is associated with selective nerve terminal damage in both CNS and PNS (LoPachin et al., 2003). A growing body of evidence indicates that the nerve terminal is a primary site of ACR action and the inhibition of corresponding membrane fusion processes impairs neurotransmitter release and promotes eventual degeneration (LoPachin, 2004).

Table 1. Neurotoxic effects of acrylamide and ethanol in experimental models

Neurotoxin and its effects	Model	Reference
Acrylamide		
Locomotor dysfunction; induction of oxidative stress, mitochondrial dysfunction	*Drosophila melanogaster*	Prasad and Muralidhara, 2012
Development of neuropathic signs; no alteration in the levels of Nrf 2, Keap1 and ARE protein products in brain and liver	Rats	Zhang et al., 2012
Pre and perinatal exposure results in oxidative stress in cerebellum; loss of Purkinje cells evident	Rat	Allam et al., 2011
Increased expression of Cox-2 and NOS	MCF-12-A	Cook et al., 2011
Toxic effects in cerebellum; increased expression of GABA, astrocytes hypertrophy	Rats	Shi et al, 2011
Increased ROS, apoptosis and DNA fragmentation, decreased cell viability	PC 12	Mehri et al., 2011
Reduced proliferation of neuronal progenitor cells; apoptosis	Mouse	Park et al., 2010
Increased levels of ROS and 3-nitrotyrosine	PC 12	Tareke et al., 2009
Oxidative stress in brain, spinal cord and sciatic nerve	Rat	Zhu et al., 2008
Ethanol		
Thermal and mechanical hyperalgesia, oxidative and nitrosative stress mediated cytokine signaling	Rats	Tiwari et al., 2012
Mechanical and thermal hyperalgesia, mechano-tactile allodynia, motor incoordination, reduction in motor nerve conduction velocity; elevation in nitric oxide, malondialdehyde, myeloperoxidase content along with reduction in membrane bound ATPase activity in sciatic nerve	Rats	Raygude et al., 2012
Hyperalgesia and involvement of PK C signaling	Rats	Alvarez et al., 2011
Enhances antiretroviral painful peripheral neuropathy by mitochondrial mechanisms	Rats	Ferrari and Levine, 2010
Mechanical hyperalgesia, involvement of sympatho-adrenal stress axis	Rats	Dina et al., 2008

Studies have shown that ACR forms Michael-type adducts with nucleophilic residues on proteins (LoPachin and Gavin, 2008). Oxidative stress and mitochondrial dysfunctions have been demonstrated to be key mechanisms in ACR-induced cell injury and neuropathy. Previously ACR administered to rats markedly elevated the levels of lipid peroxidation, activities of Glutathione- S-Transferase and SOD in plasma, liver, testes, kidney and brain (Yousef and El-Demerdash, 2006). In rodent models, ACR-induced sensory and motor dysfunction was associated with oxidative stress in nervous system (brain, spinal cord and sciatic nerve) (Zhu et al., 2008). Numerous mechanistic approaches in cell models have also shown the involvement of oxidative stress mechanism/s. ACR induces cytotoxic effects in PC12 cells and induces apoptotic cell death *via* elevation in the levels of ROS in mouse neuronal progenitor cells (Mehri et al., 2011; Park et al., 2010). Similarly, ACR induced significant cytotoxicity in human neuroblastoma cell line SH-SY5Y and glial cell line U-1240MG (Chen et al., 2009) and increased peroxynitrite–3 nitrotysosine generation in PC12 cells (Tareke et al., 2009). In a recent study, ACR caused increased expression of Cox-2, nitric oxide synthase in MCF-12-A, a non-tumorigenic epithelial cell line established from fibrocystic breast disease (Cook et al., 2011) ACR exposed mouse epidermal cell line JB6 P+ resulted in activation of transcription factors, AP-1 and NF-kB in (Lim et al., 2011) (Table 1).

2.4.2. Ethanol

Chronic consumption of alcohol damages nervous system and it is often asymptomatic. However the occurrence of peripheral neuropathy ranges from 10-50%. The complication of chronic alcohol consumption leads to allodynia, hyperalgesia and neuropathic pain with burning sensation, primarily in the lower extremities collectively called as alcohol induced polyneuropathy (Naik et al., 2006). Similar behavioural characteristics have been reported in animal models (Zimmermann, 2001). Chronic ethanol administration results in oxidative damage to macromolecules (Mansouri et al., 2001; McDonough 2003), activation of cytokines, such as TNF-α, interleukin-1b (IL-1b) (Zelenka et al., 2005) and these factors together result in the development of neuropathic condition. Alcohol consumption leads to axonal neuropathy characterized by Wallerian degeneration of the axons and a reduction in the myelination of nerve fibres (Yerdalen et al., 2008). In addition, the metabolites of ethanol such as acetaldehyde have a direct neurotoxic effect (Koike et al., 2003). Alcohol administration in animal models showed direct toxic effect on peripheral nerve and spinal cord (Narita et al., 2007; Corsetti et al., 1998). In rats, chronic alcohol administration led to allodynia/ hyperalgesia and stress accompanied with activation of inflammatory process and protein kinase signalling (Dina et al., 2000, 2008). Also, alcohol induced production of ROS leads to the damage of nerves such as the sciatic nerve, further hampering neurotransmission (Narita et al., 2007; Bosch-Morell et al., 1998) (Table 1).

2.4.3. Other Chemicals

Many chemical neurotoxins have been employed to study neurodegenerative diseases (NDD). Various adaptive mechanisms, dose dependent response to these chemicals have been extensively studied in various models to understand the pathophysiology of these NDD in humans. Some of these chemicals are discussed below:

2.4.3.1. Rotenone (ROT)

Rotenone, a commonly used natural pesticide, is a specific inhibitor of complex I (NADH - cytochrome c reductase) of the ETC. Being lipophilic it easily crosses the blood brain barrier and accumulates in subcellular organelles like Mt (Sherer et al., 2003a). It reproduces key behavioral symptoms and neuropathological features of PD (Betarbet and Greenamyre, 2007). ROT induced alterations include apomorphine-mediated rotational motor problems, nigrostriatal dopaminergic degeneration, elevated oxidative stress and changes in the brain reminiscent of Lewy bodies; selective depletion of dopamine from neurons of the substantia nigra and accumulation of α synuclein (Betarbet et al., 2000; Cannon et al., 2009; Sherer et al., 2003b). In rats, ROT administration (0.5 mg/ kg bw/ d for 7 days) caused elevation in the hydroperoxide and malondialdehyde (MDA, marker for lipid peroxidation) levels and reduced the activities of antioxidant/ detoxifying enzymes in different regions of brain and produced PD like symptoms (Denny Joseph and Muralidhara, 2012).

2.4.3.2. MPTP

MPTP (1-methyl-4-phenyl-1,2,3,6-tetrahydropyridine) is a neurotoxin that selectively destroys nigrostriatal dopaminergic neurons in humans, primates and rodents (Selvaraj et al., 2009). Being highly lipophilic, it crosses the blood brain barrier and in brain it gets converted to its toxic metabolite MPP^+ (1-methyl-4-phenylpyridinium) which selectively accumulates in dopaminergic neurons. MPP^+ interferes with mitochondrial metabolism *via* inhibition of ETC, MMP reduction and disturbances in Ca^{2+} homeostasis. This in turn triggers further oxidative stress and apoptosis, ultimately leading to cell death (Hacioglu et al., 2012). Because of loss of dopaminergic neurons, experimental animals administered with MPTP display PD like symptoms such as locomotor dysfunction or co-ordination difficulties.

2.4.3.3. 3-Nitropropionic Acid

3-Nitropropionic acid a well-known neurotoxin found in fungi, causes significant neurotoxicity in both animals and humans. It is widely used to model Huntington's disease in rodents. It causes neuronal damage and targets specially the striatum causing many neurochemical and histological features of Huntington's disease (Beal et al, 1993). It acts on the mitochondria and specifically inhibits the enzyme succinate dehydrogenase that acts in TCA cycle and the ETC at complex II (Huang et al, 2006). When administered to rats, it causes elevation in the levels of ROS and MDA in brain regions. This also resulted in elevation of free fatty acids along with increase in free radical generation.

2.4.3.4. Paraquat

Paraquat is a commonly used herbicide, whose chemical structure of is similar to the structure of the known dopaminergic neurotoxin, MPP^+ (Castello et al, 2007; Dinis-Oliveira et al, 2006). It belongs to the class of redox cycling compounds capable of inducing mitochondrial damage, generate ROS and oxidative stress in brain (Castello et al, 2007; Dinis-Oliveira et al, 2006). Administration of paraquat to mice or rats results in induction of oxidative stress, specific loss of dopaminergic neurons of the substantia nigra along with locomotor dysfunction (Somayajulu-Nitu et al., 2009). Accumulation of α synuclein (Lewy bodies) has also been reported among paraquat administered rodents, which a salient feature of PD (Manning-Bog et al., 2002).

2.4.4. Metals

2.4.4.1. Iron (Fe)

Fe is a transition metal that commonly participates in redox reactions. Fe has important role in oxygen transportation, myelin synthesis, neurotransmitter production and as a part of ETC. It is implicated in oxidative damage due to its participation in the Fenton reaction that results in the formation of hydroxyl free radical. Since most of the Fe *in vivo* is in the bound form, there is a check on its ability to participate in Fenton reaction. Fe is found throughout the brain and Fe containing proteins include cytochromes, ferritin, aconitase, non-heme iron, cytochrome P450 and many enzymes (Burdo and Connor, 2003; Zecca et al., 2004). Several brain areas have high Fe content such as substantia nigra. Catalytic iron (and copper) released due to brain damage can persist because cerebro spinal fluid has little or no Fe binding capacity. Presence of Fe (and copper) accelerates lipid peroxidation in brain *via* Fenton reaction. Hydroperoxides can further inactivate sensitive enzymes and/ or react with Fe to form hydroxyl radicals. Fe also brings about pro-oxidant activity of vitamin C, leading to further oxidative damage in the vicinity. Fe clearly induces or promotes neurotoxicity *via* free radical formation, enhanced lipid peroxidation, oxidation of proteins and alteration in the cellular and organelle membrane, followed by apoptosis (James and Steven, 2009). Fe accumulation and its involvement in redox reactions are implicated in various disease conditions such as cerulsoplasmin deficiency and neurodegenerative disorders such as Parkinson's disease (PD) (Halliwell, 2006; Asaeda et al., 2005). High cholesterol levels induced lipid peroxidation of blood brain barrier or any other type of damage to the blood brain barrier leads to increased Fe entry into the brain. A recent study reported that increased Fe levels in brain due to microbleeds that is common in AD, enhanced neurotoxic effects of amyloid b in rats (Wang et al., 2012).

2.4.4.2. Lead (Pb)

Humans are exposed to Pb *via* paint, water and food containers. Along with various other vital organs, Pb has a specific effect on CNS. The neurotoxic effects of Pb among children and adults are evident by symptoms like irritability, head-ache decreased attention span and cognitive impairment/s. Persistent childhood exposure to Pb leads to impulsiveness, inability to follow directions, decreased play activity, lowered IQ and poor attentiveness and neuroinflammation in mice brain (Papanikolaou et al., 2005; Kaston-Jolly et al., 2012). Chronic exposure to Pb results in abdominal pain or cramping, nausea/ vomiting, short term memory loss, depression, in coordination, numbness, tingling in extremities and inability to concentrate (Patrick, 2006). Pb exposure toxicity in the radial and peroneal nerves especially as seen among painters, results in wrist drop and foot drop; commonly the upper extremities are affected more and early than the lower extremities. Rats exposed to low levels of Pb, results in increased lipid peroxidation in brain (El-Sokkary et al., 2003). One of the important toxic effects of Pb is that it mimics and competes with calcium in bone formation as well several enzymatic and non-enzymatic reactions such as the active site of phosphokinsase C. This process interferes with calcium homeostasis, leading to alteration in the mitochondrial structure/ function and neuronal functions especially neuronal signalling / reactions governed by calcium (Bressler and Goldstein, 1991). Pb is also a developmental neurotoxin as it interferes with the maturation of blood brain barrier during foetal development and exerts toxic effects in immature astroglia and interferes with myelin formation thereby affecting neuronal plasticity. In addition, 'rupture' of the blood brain barrier causes the entry of various proteins and foreign bodies/ toxins in to the CNS, resulting in edema, increased intracranial pressure encephalopathy (Holtzman et al., 1984). Pb toxicity causes aberrations in both CNS and PNS. Recent findings suggest that Pb induced adverse outcomes, such as reduced intelligence quotient and academic deficits, occur at blood levels even below10 μg/dl (David, 2008).

2.4.4.3. Aluminium (Al)

Al is the most widely distributed metal in the environment and humans are exposed to Al *via* air, food, beverages, Al containing drugs and water. The other sources of Al are food additives, containers, cookware, utensils and food wrappings. While Al has no known physiological role, chronic exposure of animals to Al results in behavioural, neuropathological and neurochemical changes. Common symptoms of Al neurotoxicity are deficits of learning and behavioural functions (Kumar and Gill, 2009). In rodents, Al exposure leads to oxidative stress and oxidative damage to vital proteins in different regions of brain (Jyoti et al., 2007). Al has been implicated in Alzheimer's disease (AD) and Al content in AD brain is known to be higher (Bharathi et al., 2008). In AD, Al in brain and spinal cord is known to interfere with neurofilament protein expression and phosphorylation (Savory et al., 2006). In addition, Al induced DNA damage and apoptosis in various models (Lima et al., 2007; Fu et al., 2003; Ghribi et al., 2001). Al interferes with the expression of important genes; while it activates monoamine oxidase and Cox-2, it decreased the expression of NGF, BDNF and TNF α (Huh et al., 2005; Johnson and Sharma, 2003; Jun-Qing et al., 2006).

2.5. Hereditary Neuropathy

Hereditary neuropathy (HN) predominantly affects peripheral nerves and is associated with symptoms of peripheral nerve damage. Other hereditary neuropathies affect both the central and peripheral nervous systems and in some cases, other organs. This has led to the difficulty in diagnosing such patients, as symptoms related to peripheral neuropathy may be overshadowed by other manifestations of the disease. Due to variation in the phenotypes a generalized classification of HN has been tough, however, two broad groups are (i) hereditary motor-sensory neuropathy (HMSN) and (ii) hereditary sensory-autonomic neuropathy (HSAN), and these are further subdivided based on specificity of symptoms and dysfunction. Some of the examples of HN are discussed below:

2.5.1. Charcot-Marie-Tooth disease (CMT)

CMT is the commonest type of hereditary neuropathy, usually is synonym for HMSN. It is a neuromuscular disorder affecting at least one in 2500 US citizens (Szigeti and Lupski, 2011). The main features of CMT are a combination of lower motor neuron type motor deficits and sensory loss with muscle wasting and weakness. It is a combination of clinically heterogeneous group of disorders caused by mutations in the genes of the peripheral nerves that is, the nerves spreading all over the body – the sensory organs in the limbs. Thirty six loci and more than 30 genes are involved in CMT disease manifestation (Reilly et al., 2011). These genes take part in different pathways involving various processes such as myelination, Schwann cell differentiation, radial and axonal transport, signal transduction, mitochondrial function, endosome protein translation and single stranded DNA break repair. The clinical symptoms include weakness of the foot and lower leg muscles, which may result in foot drop and a high-stepped gait with frequent tripping or falls. Foot deformities, such as high arches and hammertoes (a condition in which the middle joint of a toe bends upwards), are also characteristic due to weakness of the small muscles in the feet. Later in the disease, weakness and muscle atrophy may occur in the hands, resulting in difficulty with fine motor skills. Some patients experience pain, which can range from mild to severe (Reilly et al., 2011; Szigeti and Lupski, 2011). Clinical phenotypes are established based on the age of onset, neurophysiological findings, electrophysiological studies, peripheral nerve MRI as well as sural nerve and skin biopsies. Clinical phenotypes include Charcot–Marie–Tooth disease type 1 (CMT1), Charcot–Marie–Tooth disease type 2 (CMT2), Dejerine–Sottas neuropathy (DSN), congenital hypomyelinating neuropathy (CHN) and Roussy– Levy syndrome (RLS). In general, weakness starts distally in the feet and progresses proximally in an ascending pattern. Neuropathic bony deformities develop including *pes cavus* (high-arched feet) and hammer toes. With further progression, muscle stretch reflexes disappear in ankles and later in upper limbs. Mild sensory loss to pain, temperature or vibration sensation in the legs is consistent with the phenotype.

2.5.2. Hereditary Sensory and Autonomic Neuropathy (HSAN)

HSANs occur rarely and are evident in early childhood. The major feature of the HSANs is loss of large myelinated and unmyelinated fibers (Houlden et al., 2006). Loss of sense in general and absence of pain in particular with self-mutilation are characteristics of features of this syndrome. Because of loss of pain, children start traumatizing themselves at an early age (2-3 years old) as a result of this, ulcers and stress fractures develop in fingers and toes. Besides environmental factors, children get injured by their own teeth, show lingual / labial injuries with torn-off lips and tongue (Erdem et al., 2000). Using of an oral shield prevents the biting and thus traumatization of the tissue can be prevented. A recent study reported that oral supplementation of L-Serine reduces the production of neurotoxic deoxysphingolipids in mice and humans with hereditary sensory autonomic neuropathy type 1 (Garofalo et al., 2011). Although HSANs have been classified in to five sub-types, patients (children) cannot be categorized with clear-cut distinction in to any of these groups. However the current understanding about the genetic abnormalities is scarce. The five sub-types are HSAN type 1, HSAN type 2, HSAN type 3, HSAN type 4 and HSAN type 5 (Erdem et al., 2000).

2.5.3. Hereditary Neuropathy with Liability to Pressure Palsies (HNPP)

HNPP is a sensory and motor multiple mononeuropathy with autosomal dominant transmission (Li et al., 2004; Ohyagi et al., 2012). In HNPP, minor or major and recurrent compression of peripheral nerves at multiple sites is evidence by the electrophysiological studies. There is usually no difficulty distinguishing these patients from CMT patients as they typically present with transient, recurrent episodes of focal weakness or sensory loss, usually precipitated by pressure, in the distribution of individual nerves or plexus (Li et al., 2004; Szigeti and Lupski, 2009). But with time as the clinical condition progresses, patients exhibit symptoms that overlap with CMT. Electrophysiological findings include mildly slowed NCV, increased distal motor latencies and conduction blocks (Li et al., 2002; Reilly et al., 2011). Some of the nerves commonly involved are peroneal nerve at fibular head and ulnar nerve at elbow and the neuro-pathological hallmark is sausage-like thickening of myelin sheaths (Szigeti and Lupski, 2009; Ohyagi et al., 2012).

2.6. Traumatic Injuries, Post-Operative, Spinal Cord Injury

There are few other reasons which may also result in nervous tissue damage or neuronal dysfunction, for eg. traumatic injuries, post-operative/ surgery, postherpetic neuroglia and spinal cord injury.

3. Current Status of Therapies Available

Conventional analgesics such as non-steroidal anti-inflammatory drugs and opioids are ineffective against neuropathic pain. Tricyclic antidepressants (eg. amitriptyline, nortriptyline, imipramine) and anticonvulsants (eg, phenytoin, gabapentin, carbamazepin) have been reported to produce anti-allodynic effects in neuropathy. Some of the drugs that have been approved and licensed are: Lyrica (pregabalin), Cymbalta (duloxetine), Neurontin (gabapentin), Lidoderm (5% lidocaine patch) and Qutenza (8% capsaicin patch) (Garg and Adams, 2012). However, their full clinical exploitation is limited due to a wide spectrum of adverse effects associated with their clinical use and none of the medications, assessed in randomized controlled studies conducted has been found to effective against neuropathic pain (Choi and Mi-won, 2011) (Table 2).

Table 2. Common drugs prescribed for management of symptoms of neuropathy / neuropathic pain

Drug	Effects	Side effects
Amitryptyline	Antidepressant, nerve pain relief	Dry mouth, sedation, dizziness, confusion, hypotension, constipation, urinary retention, blurred vision
Gabapentin	Anticonvulsant, nerve pain relief	Dry mouth, sedation, dizziness, constipation, insomnia, sexual dysfunction, diarrhea
Tramadol	Analgesic, antidepressant	Nausea, sedation, constipation, headache, dry mouth, urinary retention, confusion, tremor, seizures; lack of tolerance in long term use and among elderly
Capsaicin and lidocaine	Pain relief (topical application)	Burning, itching
α Lipoic acid	Limited improvement in neuropathic symptoms; improves nerve function	Benefits under long term usage in DN

4. Spices and Medicinal Plants – Promising Therapeutic Tool

4.1. Spice Active Principles (SAPs)

Spices are in use since time immemorial as food adjuncts and in traditional systems of medicine such as Ayurveda. Spices can be leaf (bay leaf), buds (cloves), rhizomes (turmeric), seeds (cumin), bark (cinnamon), and stigma of flower (saffron) of a plant. Spices are known to possess various bio-actives in addition to varied quantities of essential oils, proteins, sugars, fibres, minerals and pigments. These bioactive compounds are mostly phenolic acids, flavonoids, sterols or terpenes. The active principles of some of the spices are: curcumin from turmeric, eugenol/ isoeugenol from clove, cinnamaldehyde from cinnamon, capsaicin from chillies, crocin from saffron, piperine from pepper, gingerol from ginger and geraniol from nutmeg, coriander and ginger. In the recent past, there are several reports which provide evidence for the various health benefits of different SAPs (Srinivasan, 2012). Their role in enhancing neuronal health under disease condition or neurotoxin insults is extensively addressed in various experimental models. Most of the SAPs are very good antioxidants and possess anti-inflammatory activity.

4.1.1. Curcumin

Curcumin (deferuloylmethane) is one of the widely studied SAPs, which exhibits various biological and pharmacological properties. Studies with curcumin have revealed the neuroprotective function in neurodegenerative disorders such as AD and PD and in metabolic diseases such as DN. Chronic curcumin treatment of diabetic mice (induced by STZ) significantly attenuated thermal hyperalgesia and hot plate latencies. Curcumin also inhibited the TNF-α and NO release in brain (Sharma et al., 2006). STZ induced diabetic rats exhibited decreased gene expression of insulin receptor, Muscarinic M1, SOD, choline acetyl transferase was elevated by curcumin treatment. The elevated expression of genes such as acetyl choline receptor and acetylcholinesterase in cerebral cortex of the diabetic rats was significantly reduced by curcumin treatment (Kumar et al., 2011). Chronic curcumin treatment to old rats resulted in decrease in lipid peroxidation and lipofuscin contents with a concomitant elevation in the activities of SOD, glutathione peroxidase and Na^+ K^+ ATPase in brain regions (Bala et al., 2006). Treatment of BSO exposed dopaminergic neuronal cells as well as mice with curcumin restored the depleted GSH levels, protected against protein oxidation and preserved complex I activity (Jagatha et al., 2008). Curcumin treatment to AD model mice, significantly improved the memory ability in step-through test and attenuated the neuropathological changes in the hippocampus. Curcumin also inhibited apoptosis by increasing the Bcl-2 levels (no change in Bax) and this perhaps resulted in reduced neuropathological changes (Rui et al., 2008). Curcumin attenuated $AlCl_3$ induced apoptosis in PC 12 cells (Rui et al., 2008). Curcumin and its metabolite tetrahydrocurcumin inhibited monoamine oxidase-B in MPTP model of PD in mice (Rajeswari and Sebesan, 2008). *In vitro* studies showed inhibition of Aβ fibrils formation from Aβ 40 and Aβ 42 by curcumin in a dose dependent manner (Ono et al., 2004). Recently, curcumin was shown to prevent oxidative/ nitrosative stress and inflammatory reactions along with behavioural aberrations induced in a rat model of alcohol neuropathy (Kandhare et al., 2012). Our recent findings showed that curcumin rescued *Drosophila* flies against ACR induced locomotor dysfunction and lethality in a co-exposure paradigm (Prasad and Muralidhara, Unpublished).

4.1.2. Eugenol and Isoeugenol

The major active components in cloves viz., eugenol, isoeugenol and methyleugenol are known to possess antioxidant, anti-septic, local anaesthetic, and anti-inflammatory properties (Hidalgo and Rosa, 2009). Both eugenol and isoeugenol belong to the o-methoxy phenols class of molecules (Choi et al., 2007; Hidalgo and Rosa, 2009). In cell models eugenol is shown to enhance the activities of several glutathione –related enzymes and protect primary neuronal cells from excitotoxic and oxidative injury (Wie, 2006; Wie et al., 1997; Mahapatra et al., 2009). Eugenol inhibited the activity of 5 – lipoxygenase activity in human polymorphonuclear leukocytes (Prasad et al., 2004). Previously, isoeugenol was shown to suppress inducible nitric oxide synthase expression and nitric oxide mediated inflammatory reaction (Mahapatra et al., 2009). Eugenol derivatives eugenolol and glyceryl-isoeugenol suppressed LPS induced expression of iNOS in mouse macrophages by down regulating NF κB and AP-1, through inhibition of MAP kinases and Akt/IκB α signalling pathways (Yeh et al., 2011). Bis-eugenol, a eugenol ortho-dimer and dehydrodiisoeugenol (an isoeugenol dimer) inhibited LPS -induced NFκB activation and inflammatory response in macrophages (Murakami et al., 2003; Murakami et al., 2005). Further, both eugenol and isoeugenol were shown to significantly inhibit lipid peroxidation and prevent depletion in dopamine levels in experimental animals (Kabuto et al., 2007). In a recent study, Eug pre-treatment was shown to prevent increase in Cox-2 gene expression and arrest thioacetamide-induced DNA strand breaks in rats (Baskaran et al., 2010). A body of evidence have demonstrated the inhibitory action of eugenol on AChE activity, suggesting its potential therapeutic role in neurodegenerative disease such as AD (Dohi et al., 2009). Eugenol treated snails exhibited inhibition of AChE activity in the nervous tissue (Kumar et al., 2009). Earlier reports have shown that eugenol provided significant preservation of neuronal cells in the CA1 region of hippocampus after 5 min occlusion of the carotid arteries in the Mongolian gerbils (Won et al., 1998). It also rendered protection to neuronal cells from excitotoxic and oxidative injury in primary cortical cultures against N-methyl-d-aspartate (NMDA)-, oxygen-glucose deprivation-, and xanthine/xanthine oxidase- induced neurotoxicity in primary murine cortical cultures (Wie et al., 1997). Recent findings from our laboratory have clearly shown that eugenol/ isoeugenol enriched diet provides marked protection against ACR induced locomotor deficits, lethality and oxidative stress in adult Drosophila in a co-exposure paradigm (Prasad and Muralidhara, 2012). Further, treatment of acrylamide administered adult rats with eugenol/ isoeugenol resulted in marked improvement in gait score and behavioural tests for sensory and motor function (Prasad and Muralidhara, 2013). Both the actives markedly ameliorated the elevation of oxidative stress markers induced by ACR in different brain regions and sciatic nerve (Prasad and Muralidhara, 2013). Other functions exhibited by eugenol/ isoeugenol include rubefacient, carminative and anti-flatulent properties (Hidalgo and Rosa, 2009).

4.2. Medicinal Plants

Herbal extracts and or/ their bioactive components have revived interest among the research community for their immense role in modulating a disorder or disease condition related to any of our vital organs and nervous system is no exception. While Ayurveda (system of Indian traditional medicine) has a list of about 2,000 plant species with medicinal value, Chinese Pharmacopaoeia lists over 5,700 medicinal plants. In this section, we have discussed the neuroprotective properties of selected plants.

4.2.1. Bacopa Monneri (BM)

(Family: Scrophulariaceae), BM has been used by Ayurvedic medical practitioners in India for almost 3000 years and is classified as medhya rasayana, (a drug for improving memory and intellect). BM is known as 'brahmi', derived from the word 'Brahma' which in Hindu mythology refers to the 'the Supreme creator'; this is perhaps to emphasize its role in enhancing brain (a centre for knowledge and creative activity) health and function. BM is locally known as 'jalanimba and jalabrahmi' also (Chopra et al., 1956). BM Linn. (syn. *Herpestis monnieri* Linn.), is an annual creeping plant found throughout India, Nepal, Srilanka, China and Taiwan. BM has been extensively being used as nerve/ brain tonic to enhance memory, as antianxiety, antiepileptic agent and also used against cognitive dysfunctions (Husain et al., 2007). Actives ingredients reported in BM are: alkaloids (nicotine and herpestine), characteristic saponins ('bacosides' – bacoside A, bacoside B,) D-mannitol, potassium salts, phenylethanoid glycosides and different flavonoids (Chopra et al., 1956). Both BM extracts and purified ingredients such as bacosides have been extensively investigated for their neuro-pharmacoligical effects. BM possesses anti-inflammatory, analgesic, antipyretic, sedative and antioxidant properties. The other functions include its ability to chelate metal ions, reduce nitrosative stress, interference with neurotransmitter action or metabolism such as dopamine or cholinergic function. Several studies in different models of AD, PD or metal toxicity such as Al has revealed the protective effects of BM (Holocomb et al., 2006; Jyoti and Sharma, 2006; Hosamani and Muralidhara, 2010). In AD transgenic mouse model, BM extract inhibited multiple components of beta amyloid induced oxidative stress pathway that can contribute to AD pathology and reduced beta-amyloid levels in the brain (Dhanasekharan et al., 2007). Methanol extract of BM provided effective protection against DNA cleavage in human fibroblasts induced by HP-UV photolysis and in rat astrocytes induced by nitric oxide donor (Russo et al., 2003a, b).

Studies from our laboratory have shown the antioxidant potential and neuroprotective effects in Drosophila models of PD (Hosamani and Muralidhara, 2010; Hosamani and Muralidhara, 2009). BM leaf extract enhanced antioxidant defenses in both cytosol and mitochondria, caused diminution in the levels of MDA, ROS, hydroperoxides, protein carbonyls along with significant reduction in AChE activity in different regions of brain of prepubertal mice (Shinomol et al., 2012). BM enriched diet offered significant protection against chemical neurotoxins such as Rotenone and 3-nitroproionic acid (Shinomol et al., 2012a, b). BM also up-regulates adaptive mechanisms by modulating Hsp 70 expression and downstream antioxidant enzymes (Chowdhuri et al., 2002). BM extract reversed the cognitive deficits induced by intra-cerebroventricularly administered colchicines and injection of ibotenic acid into the nucleus basalis magnocellularis in rats; depletion of acetylcholine, activity of choline acetylase and muscarinic cholinergic receptor were reduced in frontal cortex and hippocampus (Bhattacharya et al., 1999). Recently in mice, BM exhibited ameliorative effect on amnesic condition induced by diazepam involving behavioural-molecular partitioning (Saraf et al., 2008). Because of such diverse functions and the ability of BM to alleviate neuro-related disease condition, it is rightly accepted to possess nootropic properties. In several clinical studies BM either given alone or in combination has indicated positive results with respect to cognitive function, anxiety, attention disorder, nervousness, palpitation, insomnia, etc., without any clinically significant side effects (Singh and Singh, 1980; Maher et al., 2002).

4.2.2. Withania Somnifera (WS)

(Family: Solanaceae), commonly called 'ashwagandha', is also referred to as 'Indian ginseng'. The major constituents of WS are alkaloids (isopelletierine and anaferine), saponins (sitoindosides) and the C28 steroidal lactones- withanolides (Elsakka et al., 1990; Misra et al., 2005, 2008). Other constituents include glycosides, starch and a variety of amino acids such as aspartic acid, proline, tryptophan and alanine. WS is known for its diverse therapeutic applications including nootropic effects (Gokul et al., 2012). WS extract preferentially acts on the cortical and basal forebrain cholinergic signal transduction cascade – contributing to cognition enhancing events (Schliebs et al., 1997). WS displays anti-convulsant properties in both acute and chronic models of epilepsy induced by pentylenetetrazol (PTZ) (Kulkarni and Verma, 1993; Kulkarni et al., 1993). Pre-treatment with WS root extract significantly attenuated 3-NPA induced cognitive deficits (Kumar and Kumar, 2008). WS methanolic extract significantly improved behavioural responses such as active avoidance activity and spontaneous locomotory response in 'mice models and also abrogated diabetes induced memory degeneration (Parihar et al., 2004). WS withanoloids exhibit inhibitory action on the activities of both acetyl- and butyl-cholinesterase enzymes and calcium imbalance, thus influencing the cognitive functions (Choudhary et al., 2005). WS extract also increased the cortical muscarinic acetylcholine receptor capacity both in animal models and humans (Schliebs et al., 1997). Pre-treatment with WS glycowithanolides produced a dose-related reversal of the stress effects in a chronic foot shock-induced stress model (Bhattacharya et al., 2001). Effect of WS on CNS includes significant regeneration of both axons and dendrites in addition to the reconstruction of pre- and post-synapses in neurons, neurite outgrowth in rat cortical neurons, attenuation of AD induced neurodegeneration and rescued neuronal death by promoting axonal growth, increased peripheral myelin level and facilitated locomotion in spinal cord injured mice (Gokul et al., 2012; Kuboyama et al., 2006; 2005). Recently, we have demonstrated the prophylactic neuroprotective effect of WS in mice, against rotenone induced oxidative impairments and mitochondrial dysfunctions in striatum and cerebellum which has implications for the potential of WS against PD (Manjunath and Muralidhara, 2013). Clinical investigations with WS root extracts suggest that it exerts significant anti-aging effect in normal healthy but aged subjects. WS also exhibited hypoglycemic, diuretic and hypocholesterolemic effect (Hemalatha et al., 2004; Andallu and Radhika, 2000).

4.2.3. Centella Asiatica (CA)

(Family: Apiaceae): The prominent group of biologically active compounds isolated from CA are terpenes. The extract has certain bioactive terpene acids such as Asiatic acid, madecassic acid and their respective glycoside, asiaticoside and madecassoside.

Several derivatives of asiaticoside and Asiatic acid were found to be protective effect against β- amyloid induced neurotoxicity and associated dementia (Howes et al., 2003). CA finds application in wound healing, immune booster, memory improvement, better circulation, sedative/ mental calmness, anti-stress and anti-anxiety (Husain et al., 2007) and skin disorders such as psoriasis, eczema, leprosy, healing of wound, burns, skin and vein ulcers (Kartnig, 1998; Viala et al., 1977). CA prevented age related changes in antioxidant defense system, lipid peroxidation and protein carbonyl content (Subathra et al., 2005).

Oral administration of CA resulted in decreased PTZ kindled seizures and exhibited improvement in learning deficit induced by PTZ kindling in rats. Our recent studies in mice, showed prophylaxis CA aqueous extract conferred protection against 3-NPA induced oxidative stress- elevation in ROS, MDA, HP and PCs, in different regions of brain (Shinomol et al., 2010). CA leaf juice improved passive avoidance task in albino rats (Nalini et al., 1992).

The abnormal activity of phospholipase A_2 (both Ca^{2+} independent and cytosolic) in CNS during epilepsy, stroke, multiple sclerosis and neuropsychiatric disorders was inhibited by CA extract in a dose dependent manner (Barbosa et al., 2008). CA extract administration (two months) enhanced working memory and improved self-rated mood among healthy elderly volunteers in a randomized, placebo-controlled double-blind study (Wattanathorn et al., 2008).

4.2.4. Ginkgo Biloba (Gb)

Gb and its extract, is an oriental herbal Chinese medicine with various pharmacological effects. The standardized Gb extract EGb 761 is one of the most widely used herbal medicine for dementia and cognitive impairments, and remains one of the best evaluated and characterised extracts.

Table 3. Neuroprotective properties of melatonin and Resveratrol in experimental models of neuronal dysfunction

Active/ Property	Model	Reference
Melatonin		
Ameliorates locomotor deficit; lowers oxidative stress and mitochondrial dysfunction in MPTP model of PD	Mice	Patki and Lau, 2011
Attenuated thermal hyperalgesia, cold allodynia and oxidative damage in chronic constriction injury of sciatic nerve	Rats	Kumar et al., 2011
Improved nerve conduction velocity and nerve blood flow; reduced NFκB and cytokines in sciatic nerve in DN (STZ) model	Rats	Negi et al., 2011
Reduction in oxidative stress markers and increase in monoamine levels in acrylamide model of neuropathy	Rats	Ahmed et al., 2010
Resveratrol		
Protected the cells from ethanol induced neurotoxicity at 0.1-10 μM concentrations in primary cultures	Rat astrocytes	Gonthier et al., 2012
Decreased NFκB, lipid peroxidation; improved nerve conduction velocity and nerve blood flow in DN (STZ) model	Rats	Kumar and Sharma, 2010
Improved nerve conduction velocity and reduced axonal degeneration in sciatic nerve in 3-Nitropropionic acid induced peripheral neuropathy	Rats	Binienda et al., 2010
Neuroprotective effects by modulating mitochondrial dysfunctions in cerebral ischemia (Middle cerebral artery occlusion)	Rats	Yousuf et al., 2009

Gb extract consists of flavonoid glycosides terpene lactone (ginkgolides) and other organic acids. Gb extract is widely used in treating vascular diseases such as peripheral circulation diseases and cerebral insufficiency because of its favourable effect on blood circulation (Kleijnen and Knipschild, 1992). In the optic nerve crush model of rats, supplementation with Gb extract exhibited higher survival rate of retinal ganglion cells (Ma et al., 2010). Treatment with Gb to male volunteers exhibited significant optic nerve head blood flow (Wimpissinger et al., 2007). Gb extract EGb 761 alone and in combination with selenium rendered protection to rats with ischemia and reperfusion injury in brain by suppression of oxidative stress pathways (Erbil et al., 2008).

4.3. Others

Recently we have reported from our lab the neuromodulatory potential of aqueous extract of *Selaginella* in the rotenone model of PD in Drosophila (Girish and Muralidhara, 2012). In a rat model of stroke, administration of resveratrol, phytophenol found in red grapes and wine, reduced ischemic damage to the brain and protected spinal cord neurons against ischemic injury (Huang et al., 2001; Kaplan et al., 2005). Many *in vitro* studies have also demonstrated the neuro-protective effects of resveratrol and melatonin (Patki and Lau, 2011; Marambaud et al., 2005; Jang and Surh, 2003) (Table 4).

Table 4. List of primary cell cultures and cell lines used to study neuronal function

In vitro model	Derived from/ cell type
Primary cell cultures	
Primary cortical cells	Cerebral cortex from embryos (E16-19)- rodents
DRG neurons	Drosal root ganglion harvested from embryos (E15-rodents) or hamsters
Cerebellar granule cells	Cerebellum from 7d old pups- rodents
Schwann cell cultures	Schwann cells derived from embryos- rats
Neural crest cells	Cranial neural crest cells from embryos- rats
Cell lines	
PC 12	Rat Pheochromocytoma
SHSY5Y	Human neuroblastoma
SKNSH	Human neuroblastoma
IMS32	Rodent Schwann cells
IMR32	Human neuroblastoma
N27	Rat mesencephalon
M213-20	Rat striatum
HN2-5	Mouse hippocampal

5. Models Employed to Study Neuropathy

5.1. In Vivo Models

5.1.1. Rodent Models

Several *in vivo* models such as diabetes(STZ) induced - neuropathy , PTZ induced epilepsy, anticancer agent-induced or alcohol -induced neuropathies are used extensively (Zheng et al., 2012; Abad et al., 2011; Tiwari et al.,

2012). The amenability to genetic manipulation and/ or generation of transgenic animals is an added advantage of employing rodent models to address specific questions such as effect of toxins or bioavailability of certain nutrients/ adjuvants and metabolic disorders such as DN. There are several studies that report utilization of nerves as the focal point to study especially peripheral neuropathy. One of the most commonly used models involves unilateral loose ligation of the sciatic nerve with chromic gut sutures. Such chronic constriction injury/ sciatic nerve constriction and ligation have been useful to study neuropathophysiology *per se* as well as modulation with known drugs or phytochemicals (Xu et al., 2012; Muthuraman et al., 2008; Coronel et al., 2008). Of late, bilateral constriction has also been employed to study neuropathy as the neuropathic signs prolong unlike unilateral constriction injury that lasts for 4-5 weeks (Datta et al., 2010).

5.1.2. Drosophila Melanogaster

Drosophila has been intensively used as a model system for both basic and applied research on human neurological diseases. *Drosophila* offers several advantages for mimicking human brain diseases to understand the molecular and cellular mechanisms (Feany, 2010). The advantages of the fly model include short life span, large number of offspring, convenience of maintenance/ culturing, and many genetic tools to modulate / regulate specific gene expression and knock outs and a well-known anatomy. In the fly model, certain innate proteins and/ or human proteins can be expressed in a specific tissue and also the expression can be in a time dependent manner. Basically, a human disease related transgene is inserted downstream of a UAS (upstream activating sequence) and is expressed under the control of the yeast transcriptional activator GAL4, this is the GAL4/ UAS system.

Drosophila is widely used as a model of neurodegenerative disorders such as PD, AD and Huntington's disease. The amyloid precursor protein (APP) homolog in *Drosophila* does not contain the Aβ domain and cannot be cleaved in flies and the γ secretase, presenilins and nicastrin are conserved (Martin-Morris and White, 1990; Kopan and Goate, 2002). To model PD in fly, wild type human α-synuclein and two familial mutant forms (A30P and A53T) have been made to express in dopaminergic neurons. This results in age-related loss of dopaminergic neurons, Lewy bodies like accumulations and behavioural deficits; in addition L-DOPA supplement to these mutants resulted in improvement in behavioural deficits further authenticating the PD model in flies (Meredith et al., 2008; Pendleton et al., 2002). Fly model is also being explored to study the effect of neurotoxins, metals, metabolic diseases such as

Leigh disease, Niemann-Pick disease, tumor syndromes such as neurofibromatosis (Coulom and Birman, 2004; Hosamani and Muralidhara, 2009 ;Huang et al., 2007; Botella et al., 2009; Prasad and Muralidhara, 2012). Cisplatin induced behavioural deficits and apoptotic changes in brain of *Drosophila* resemble those seen in mammals along with cisplatin-DNA adduct formation in whole body homogenates (Podratz et al., 2011).

Recently we have demonstrated the prophylactic neuroprotective efficacy of BM extract/s, and *Selaginella* (Sanjeevini , the resurrection plant) extracts employing chemical models of PD in *D. melanogaster* (Hosamani and Muralidhara, 2009, 2010; Girish and Muralidhara, 2012). Our studies revealed a good agreement among the responses in flies and rodent model with respect to the phenotype and biochemical alterations (eg. oxidative stress, reduction in antioxidant levels and dopamine levels) under ACR exposure (Prasad and Muralidhara, 2012). Further, we also showed that various spice actives (eg. eugenol, isoeugenol, geraniol and curcumin) significantly rescued flies against ACR-induced motor deficits and lethality (Prasad and Muralidhara, 2012). Exposure of flies to high concentration of iron reduced survivability and induced motor dysfunction and polyphenols were capable of increasing the survival and climbing abilities (Jimenez-Del-Rio, et al., 2010).

5.2. *In Vitro* Models

In vitro studies can be employed to begin an investigation for the basic understanding of the pathophysiological effects. However, extrapolation of the data obtained from these studies to predict the deleterious effects *in vivo* depends in part upon the relationship between the endpoint measured and the normal biological function. Both biochemical and morphological markers can be obtained by these studies. Biochemical markers include ROS, protein carbonyls, hydroperoxides, enzymes, neurotransmitters, neurotransmitter receptors, expression of some specific genes and apoptotic factors. Morphological factors include cell survivability, cell health (for eg. clumping of cells or lifting up of adherent cell are signs of toxic effects of the insult employed) and neurite growth (initiation, length and branching of outgrowths). Though these studies always require further validation in animal models, it minimizes the use of large number of experimental animals which needs to be within ethical guidelines.

Several *in vitro* models serve as good tools to get mechanistic insights into the biochemical pathways and molecular events involved in development of neuropathy. The most extensively used *in vitro* models for studying nervous tissue related disorders - neuropathies are primary cultures, cell lines and brain slices as described below (Table 4).

5.2.1. Primary Cell Culture

Different types of primary cell cultures are in use to address various questions with regard to neuronal responses against metabolic disorder or a toxin. Cells are harvested either from the embryos (mostly rodents), or from pups (rodents) in the early post natal stage. Specifically cells can be employed from a particular tissue region such as cortex, cerebellum, dorsal root gangalia etc.

DRG are harvested from rat embryos (GD day 15) are exposed to high glucose concentration 45 mM (normal glucose concentration being 20 mM) has given mechanistic insight in to the severe effects of high glucose. Employing this model, it is understood that hyperglycemia induces oxidative stress evidenced by elevation in the ROS and lipid peroxidation levels (Vincent et al., 2005). Hyperglycemia associated mitochondrial dysfunction in DRG neurons has been evidenced in STZ diabetic rats (Kamboj and Sandhir, 2011; Rolo and Palmeira, 2006; Leinninger et al., 2006; Fernyhough et al., 2003). DRG primary cultures are also convenient systems to decipher the molecular events involved in understanding events such as apoptosis, oxidative stress mechanisms, levels of antioxidant and also levels/ expression of various growth factors (Leinninger et al., 2004; Vincent et al., 2005; Vincent et al., 2004b; Molliver et al., 1997). Other effects of hyperglycemia in DRG cultures reported are aberrations in TCA cycle enzymes and ETC (Rolo and Palmeira, 2006; Vincent et al., 2004b).

Primary cortical cell cultures prepared from rodent foetuses (GD16-19) have been employed to study the effect of toxins and neurotransmitter- release/ uptake. The effect of metals such as zinc has been studied in primary cortical neurons (Adamo et al., 2010). The effect of excess amount of glutamate, an excitatory neurotransmitter in CNS, leads to over-activation of post-synaptic neurons, leading to excessive influx of calcium, alterations in the mitochondrial membrane potential and increases in ROS (Lemus-Molina et al., 2009). Employing these models, the protective effect of *Mangifera indica* extract against glutamate-induced neurotoxicity was demonstrated (Lemus-Molina et al., 2009). Further, the modulatory effect of BM extract against beta Amyloid induced cell death was also reported (Limpeanchob et al., 2008). Cerebellar granule cell (CGC) cultures derived from the cerebellum of 7 day old rodent pups are widely employed to study various disease associated or toxin induced cytoskeletal alterations, inducers of oxidative stress and apoptosis (Yeste-Valasco et al., 2008; Zhong et al., 2007; Arakawa et al., 2006). Studies in CGC culture, has revealed mechanisms such as Ca^{2+} influx and recycling synaptic vesicle (Evans and Cousin, 2007).

5.2.2. Cell Lines

Cell lines that can be differentiated to express the characteristics of neuron such as neurite growth, have been derived from rodent and human neuroblastomas. Basically these are transformed cells (immortalized) which can be used for infinite number of passages. These cells extend neurites, synthesize neurotransmitters and their receptors thus enabling to investigate the network and electrical activity in the synapse. Neuroblastoma cell line SH-SY5Y exposed to high glucose concentrations simulates the diabetic condition and results obtained were similar to the findings in DRG neuronal cultures (Leinninger et al., 2004; Vincent et al., 2002). The cytotoxic effects of ACR have also been studied in SH-SY5Y cell line (Chen et al., 2009). ACR induced cytotoxic effect was demonstrated to be mediated *via* increased ROS in PC 12, an effect similar to findings in the rodent model (Mehri et al., 2011; Prasad and Muralidhara, 2013). Hyperglycemia induced elevation in ROS and carbonyl levels, apoptosis and altered NGF signalling were demonstrated in PC 12 cells, consistent with findings under DN conditions in vivo (Sharifi et al., 2007; Okouchi et al., 2005; Lelkes et al., 2001). IMR-32 cell line exposed to methylmercury resulted in decrease in length of the neurites. Effect of lead (Pb) in B50 cell line derived from rat neuroblastoma, resulted in decrease in neurite initiation (Audesirk et al., 1991). The effect of pesticides has also been studied in SK-N-SH and PC 12 cells (Chang et al., 2006; Nandi et al., 2006). IMS32, a Shwann cell line derived from adult mouse expresses essential transcription and neurotrophic factors. Exposure to high glucose levels in this cell line, resulted increase in sorbital content (in diabetes the activity of Aldose reductase is enhanced, thus leading to sorbital formation) along with increased expression of certain genes (Sango et al., 2006).

5.2.3. Brain Slices

Brain slice models offer unique advantages over other *in vitro* models since they largely retain the tissue architecture of the brain regions that they originated from and maintain the neuronal activities with intact functional local synaptic circuitry. Slice cultures have been well established from different regions of the brain too, such as hippocampus, striatum, cortex, spinal cord and cerebellum which enable one to study region specific effects.

5.3. Other Models

There are other models which are being employed to understand pathophysiology of various neurodegenerative disorders. Some of the recent studies including models such as *Ceanorhabditis elegans, Danio rerio* (zebra fish), *Saccharomyces cerevisiae* (yeast), have been presented in Table 5.

Table 5. Other models employed to study neuronal function under normal and disease condition

Model	Studies related to neuronal function	Reference
C. elegans (wild and transgenic)	Dominergic features of PD in terms of movement, dopamine content, dopaminergic neurons, acetylcholinesterase activity and life span employing organophosphorus insecticides	Ali and Rajini, 2012
	Motor activity loss, PD like symptoms	Zheng et al., 2011b
	Dopamine neurodegeneration, altered neuronal activity	Karpinar et al., 2009
	PD; motor and developmental defects	Kuwahara et al., 2008
	Exposure to organophate pesticides increased ROS, decreased antioxidants	Leibson and Lifshitz, 2008; Kwong, 2002
Zebra fish (*Danio rerio*)	Mutation in exon HSN2 and impaired/ loss of WNK1 function leading to HSAN type 2	Bercier et al., 2013
	Genetic manipulation (SOD1 and HSR) to mimic and understand amyotropic lateral sclerosis and motor neuron disease	Ramesh et al., 2012
	Effect of chemotherapeutics- Vincristine, bortezomib, with respect to sensory and motor axons	Khan et al., 2012
	Gene (mitofusin 2) silencing leading to developmental defects and neuromuscular alterations – model for CMT type 2A neuropathy	Vettori et al., 2011
Yeast *(Saccharomyces cerevisiae)*	Cellular toxicity of amyloid b peptides and its ability to cross the intra-cellular membranes including its effects on mitochondria	D'Angelo et al., 2013

Summary and Future Prospective

Most of the neuropathic condition with any type of causative factor, has been demonstrated to be complex and multifaceted phenomena. Spices/ medicinal plants or their actives are promising agents to cure/prevent neuropathic pathology, however, it is necessary to explore the specific molecular targets and understand their neuropharmacological action. The current understanding about the 'protective action' of these dietary supplements may be by reducing the ROS/ RNS burden along with the up-regulation of vitagenes such as members of the heat shock protein family, heme oxygenase-I and Hsp70.

Several phytochemicals have been shown to increase the resistance of neurons to various stressors by activating specific signal transduction pathways and transcription factor Nrf 2, resulting in the induction of genes regulated by the ARE. Some of them may have a direct effect on neurons or may benefit neurons indirectly by activating such processes in the astrocytes/ glial cells.

One of the advantages of employing phytochemicals is that, being natural products these chemicals generally have a less than favourable side effect profile. However, some of the clinical trials with antioxidants (eg. vitamin C) based on interference with the redox imbalance or growth factors (eg. NGF) under disease condition have yielded negative results. Few smaller interventional studies with carefully 'selected populations' have largely resulted in positive outcome.

This simply emphasizes on a detailed dose-response analysis and elucidation of the underlying molecular mechanisms of the phytochemical-induced hormesis. This would largely aid in translational research to harness their potential to prevent or alleviate neuropathy.

Acknowledgments

We wish to thank the Director, CFTRI, for his keen interest in this area of research.

Also, a grant from the Department of Science and Technology (Women Scientist Scheme WOS - A) New Delhi, Government of India (SR/ WOS-A/ LS-201/ 2008) to the first author is greatly acknowledged.

Conflict of Interest

None.

References

Abad, ANM; Nouri, MHK; Tavakkoli, F. Effect of *Salvia officinalis* hydroalcoholic extract on vincristine induced neuropathy in mice. *Chines J. of Natural Med* 2011; 9: 0354-58.

Ackerley, S; Thornhill, P; Grierson, AJ; et al. Neurofilament heavy chain side arm phosphorylation regulates axonal transport of neurofilaments. *J. Cell Biol.* 2003;161:489–495.

Adamo, AM; Zago, MP; Mackenzie, GG; et al. The role of zinc in the modulation of neuronal proliferation and apoptosis. *Neurotox Res.* 2010; 17:1-14.

Ahmed, HH; Elmegeed, GA; El-Sayeed, SM; et al. Potent neuroprotective of novel melatonin derivatives for management of central neuropathy induced by acrylamide in rats. *Eur. J. Med. Chem.* 2010;45:5452–9.

Albers, JW; Chaudry, V; Cavaletti, G; et al. Intervention for preventive neuropathy caused by cispalatin and related compound. *Cochrane Database Syst. Rev.* 2011;CD005228 .

Ali, SJ; Rajini, PS. Elicitation of dopaminergic of Parkinson's disease in *C. elegans* by monocrotophos; an organophosphorous insecticide. *CNS Neurol. Disord Drug Targets* 2012;11: 993-1000 .

Allam, A; El-Ghareeb, AA; Abdul-Hamid, M; et al. Prenatal and perinatal acrylamide disrupts the development of cerebellum in rats: Biochemical and morphological studies. *Toxicol. Ind. Health* 2011;27: 291-306.

Alvarez, P; Ferrari, LF; Levine, JD. Muscle pain in models chemotherapy induced and alcohol induced peripheral neuropathy. *Ann. Neurol.* 2012;70:101-109.

Andallu, B; Radhika; B. Hypglycemic; diuretic and hypocholesterolemic effect of winter cherry (*Withania somnifera* Dunal) root. *Indian J. Exp. Biol.* 2000;38:607-9.

Arakawa, M; Ushimaru, N; Osada, N; et al. N-acetylcysteine selectively protects cerebellar granule cells from 4-hydroxynonenal-induced cell death. *Neuroscience Res.* 2006;55:255-63.

Asaeda, M; Sakamoto, M; Kurosaki, M; et al. A non-enzymatic derived arachidonyl peroxide; 8-iso-prostaglandin F2 alpha; in cerebrospinal fluid of patients with aneurysmal subarachnoid haemorrhage participates in the pathogenesis of delayed cerebral vasospasm. *Neurosci. Lett.* 2005;373:222–225.

Audesirk, T; Audesirk, G; Ferguson, C; et al. Effects of inorganic lead on the differentiation and growth of cultured hippocampal and neuroblastoma cells. *NeuroToxicology* 1991;12:529-38.

Bala, K; Tripathy, BC, Sharma, D. Neuroprotective and anti-ageing effects of curcumin in aged rat brain regions. *Biogerontology* 2006;7:81–89.

Barbosa, NR; Pittella, F; Gattaz, WF. *Centella asiatica* water extract inhibits iPLA2 and cPLA2 activities in rat cerebellum. *Phytomedicine* 2008;17:896-900.

Baskaran, Y; Vishwanathan, P; Anuradha, CV. Investigation of antioxidant; anti-inflammatory and DNA-protective properties of eugenol in thioacetamide-induced liver injury in rats. *Toxicology* 2010;268: 204 – 212.

Beal, MF; Brouillet, E; Jenkins, BG; et al. Neurochemical and histologic characterization of striatal excitotoxin lesions produced by mitochondrial toxin 3-nitropropionic acid. *J. Neurosci.* 1993;13: 4181–4192.

Bennet, GJ. Pathophysiology and animal models of cancer related painful peripheral neuropathy. *Oncology* 2010; 15:9-12.

Bercier, V; Brustein, E; Liao, M; et al. WNK1/HSN2 Mutation in Human Peripheral Neuropathy Deregulates KCC2 Expression and Posterior Lateral Line Development in Zebrafish (*Danio rerio*). *PLoS Genet* 2013;9: e1003124.

Betarbet, R; Greenamyre, JT. Parkinson's disease: animal models. *Handb. Clin. Neurol.* 2007;83; 265-287.

Betarbet, R; Sherer, TB; MacKenzie, G; et al. Chronic systemic pesticide exposure reproduces features of Parkinson's disease. *Nat. Neurosci.* 2000;3:1301-1306.

Bharathi, PV; Govindaraju, M; Palanisamy, AP; et al. Molecular toxicity of aluminium in relation to neurodegeneration. *Indian J. Med. Res.* 2008;128:545-556.

Bhattacharya, A; Ghosal, S; Bhattacharya, SK. Anti-oxidant effect of *Withania somnifera* glycowithanolides in chronic footshock stress induced perturbations of oxidative free radical scavenging enzymes and lipid peroxidation in rat frontal cortex and striatum. *J. Ethnopharmacol.* 2001;74:1–6.

Binienda, ZK; Beaudoin, MA; Gough, B; et al. Assessment of 3-nitropropionic acid evoked peripheral neuropathy in rats: Neuroprotective effects of acetyl-L-carnitine and resveratrol. *Neurosci. Lett.* 2010;480: 117-121.

Bolanos, JP; Heales, SJR; Land, JM; et al. Effect of peroxynitrite on the mitochondrial respiratory chain: differential susceptibility of neurones and astrocytes in primary culture; *J. Neurochem.* 1995;64:1965–1972.

Bosch-Morell, F; Martínez-Soriano, F; Colell, A; et al. Chronic ethanol feeding induces cellular antioxidants decrease and oxidative stress in rat peripheral nerves. Effect of S-adenosyl-L-methionine and N-acetyl-L-cysteine. *Free Radic. Biol. Med.* 1998;25: 365–368.

Botella, JA; Bayersdorfer, F; Gmeiner, F; et al. Modelling Parkinsons's disease in Drosophila *Neuromolecular Med.* 2009;11:268–80.

Bressler, JP; Goldstein, GW. Mechanisms of lead neurotoxicity. *Biochem. Pharmacol.;* 1999;41:479-484.

Brownlee, M. Banting lecture 2004: The Pathobiology of diabetic complications; a unifying mechanism. *Diabetes* 2005;54:1615-625.

Burdo, JR; Connor, JR. Brain iron uptake and homeostatic mechanisms: an overview. *Biometals* 2003;16:63–75.

Calabrese, V., Cornelius, C., Trovato-Salinaro, A., et al. The hormetic role of dietary antioxidants in free radical-related diseases. *Current Phamaceutical Design* 2010;16:877-883.

Calkins, MJ; Delinda, A; Johnson; et al. The Nrf2/ARE pathway as a potential therapeutic target in neurodegenerative disease. *Antioxid Redox Signal* 2008;11:497-508.

Cannon, JR; Tapias, V; Na, HM; et al. A highly reproducible rotenone model of Parkinson's disease. *Neurobiol. Dis.* 2009;34:279-290.

Carozzi, VA; Canta, A; Oggioni, N; et al. Neurophysiological and neuropathological characterization of new murine models of chemotherapy-induced chronic peripheral neuropathies. *Exp. Neurol.* 2010;226: 301–309.

Castello, PR; Drechsel, DA; Patel, M. Mitochondria are a major source of paraquat-induced reactive oxygen species production in the brain. *J. Biol. Chem.* 2007;282(19):14186-14193.

Chang, PA; Wu, YJ; Li, W; et al. Effect of carbamate esters on neurite outgrowth in differentiating human SK-N-SH neuroblastoma cells. *Chem. Boil. Interact* 2006;159:65-72.

Chavko, M; Auker, CR; McCarron RM. Relationship between protein nitration and oxidation and development of hyperoxic seizures. *Nitric Oxide* 2003;9:18–23.

Chen, JH; Wu, KY; Chiu, IM; et al. Acrylamide induced astrogliotic and apoptotic responses in human astrocytoma cells. *Toxicol In vitro* 2009;23:855–61.

Childs, EA; Lyles, RH; Selnes, OA; et al. Plasma viral load and CD4 lymphocytes predict HIV associated dementia and sensory neuropathy. *Nuerology* 1999;607-613.

Choi, SZ; Mi-Won, S. A novel botanical drug for the treatment of diabetic neuropathy. *Arch. Pharm. Res.* 2011;34: 865-67.

Choi, YC; Park, KR; Lee, JH; et al. Isoeugenol suppression of inducible nitric oxide synthase expression is mediated by down regulation of NF-kB; ERK1/2 and p38 kinase. *Eur. J. Pharmacol.* 2007;576:151-159.

Choudhary, MI; Nawaz, SA; Ul-Haq, Z; et al. Withanolides a new class of natural cholinesterase inhibitors with calcium antagonistic properties. *Biochem. Biophys. Res. Commun.* 2005;334: 276–287.

Chowdhari Parmar, D; Kakkar, P; Shukla, R; et al. Antistress effects of bacosides of *Bacopa monnieri*: Modulation of Hsp70 expression; superoxide dismutase and cytochrome P450 activity in rat brain. *Phytother Res.* 2002;16:639-45.

Cleeland, CS; Farrar, JT; Hausheer, HF. Assessment of cancer related neuropathy and neuropathic pain. *The Oncologist* 2010; 15:13-18.

Cook Jr, LEL; Tareke, E; Word, B; et al. Food contaminant acrylamide increases expression of Cox-2 and nitric oxide synthase in breast epithelial cells. *Toxicol. Ind. Health* 2011;27:11–8.

Coronel, MF; Brumovsky, PR; Ho kfelt, T; et al. Differential galanin upregulation in dorsal root ganglia and spinal cord after graded single ligature nerve constriction of the rat sciatic nerve. *J. Chem. Neuroanat* 2008;35; 94-100.

Corsetti, G; Rezzani, R; Rodella, L; et al. Ultrastructural study of the alterations in spinal ganglion cells of rats chronically fed on ethanol. *Ultrastruct Pathol.* 1998;22 : 309-319.

Cory, S; Huang, DC; Adams, JM. The Bcl-2 family: roles in cell survival and oncogenesis; *Oncogene* 2003;22: 8590–8607.

Coulom, H; Birman, S. Chronic exposure to rotenone models sporadic Parkinson's disease in Drosophila melanogaster. *J. Neurosci.* 2004;24:10993-10998.

D'Angelo, F; Vignaud, H; Di Martino, J; et al. A yeast model for amyloid b aggregation exemplifies the role of membrane trafficking and PICALM in cytotoxicity. *Disease models and Mechanisms* 2013;6:206-216.

Datta, S; Chatterji, K; Kline IV, RH; et al. Behavioural and anatomical characterization of the bilateral sciatic nerve chronic constriction (bCCI) injury: correlation of anatomic changes and responses to cold stimuli. *Molecular Pain* 2010;6:7.

David, B. Very low lead exposures and children's neurodevelopment. Curr *Opinions in Pediatrics* 2008;20: 172-177.

Denny Joseph, KM; Muralidhara. Fish oil prophylaxis attenuates rotenone-induced oxidative impairments and mitochondrial dysfunctions in rat brain. *Food Chem. Toxicol.* 2012;50:1529-37.

Dhanasekaran, M; Tharakan, B; Holcomb, LA; et al. Neuroprotective mechanisms of antidementia botanical *Bacopa monniera. Phytother Res.* 2007;21:965-69.

Dina, OA; Barletta, J; Chen, X; et al. Key role for the epsilon isoform of protein kinase C in painful alcoholic neuropathy in the rat. *J. Neurosci.* 2000;20: 8614-8619.

Dina, OA; Khaser, SG; Alessandri-Haber, N; et al. Alcohol induced stress in painful alcoholic neuropathy. *Eur. J. Neurosci.* 2008;27:83-92.

Dinis-Oliveira, RJ; Remião, F; Carmo, H; et al. Paraquat exposure as an etiological factor of Parkinson's disease. *Neurotoxicology* 2006;27: 1110-1122.

Dohi, S; Terasaki, M; Makino, M. Acetylcholiesterase inhibitory activity and chemical composition of commercial essential oils. *J. Agric. Food Chem.* 2009;57:4313-18.

Eckert, A; Keil, U; Marques, CA; et al.; Mitochondrial dysfunction; apoptotic cell death; and Alzheimer's disease. *Biochem. Pharmacol. 2003*;66:1627–1634.

Elsakka, M; Grigorescu, E; Stanescu, U; et al. New data referring to chemistry of *Withania somnifera* species. *Rev. Med. Chir. Soc. Med. Nat. Iasi* 1990;94:385–387.

El-Sokkary, GH; Kamel, ES; Reiter, RJ. Prophylactic effect of melatonin in reducing lead-induced neurotoxicity in the rat. *Cell Mol. Biol. Lett.* 2003;8:461-470.

Erbil, G; Ozbal, S; Sonmez, U; et al. Neuroprotective effects of selenium and *Ginkgo biloba* extract (EGb761) against ischemia and reperfusion injury in rat brain. *Neurosciences* 2008;13:233-38.

Erdem, TL; Ozcan, I; Ilguy, D; et al. Hereditary sensory and autonomic neuropathy: review and a case report with dental implications. *J. Oral Rehabilitation* 2000;27:179-82.

Esper, RM; Pankonin, MS; Loeb, JA. Neuregulins: Versatile growth and differentiation factors in nervous system development and human disease. *Brain Research Reviews* 2006;51: 161-175.

Evans, GJO; Cousin, MA. Simultaneous monitoring of three key neuronal functions in primary neuronal cultures. *J. Neuroscience Methods* 2007;160:197-205.

Falls, DL. Neuregulins: functions; forms; and signalling strategies. *Exp. Cell Res.* 2003;284: 14–30.

Feany, MB. New approaches to the pathology and genetics of neurodegeneration *Am. J. Pathol.* 2010;176:2058–66.

Fernyhough, P; Chawdhury, SRK; Schmidt, RE. Mitochondrial stress and pathogenesis of diabetic neuropathy. *Expert Rev. Endocrinol. Metab.* 2010;5:39-49.

Fernyhough, P; Huang, TJ; Verkhratsky, A. Mechanism of mitochondrial dysfunction in diabetic sensory neuropathy. *J. Peripheral. Nerv. Syst.* 2003;8: 227-235.

Ferrari, LF; Levine, JD. Alcohol consumption enhances antiretroviral painful peripheral neuropathy by mitochondrial mechanisms. *Eur. J. Neurosci.* 2010;32:811-818.

Ferrari, S; Vento, S; Monaco, S; et al. Human immunodeficiency virus-associated peripheral neuropathies. *Mayo Clin. Proc.*; 2006;81:213-9.

Fonfria, E; Marshall, IC; Boyfield, I; et al. Amyloid b-peptide (1–42) and hydrogen peroxide-induced toxicity are mediated by TRPM2 in rat primary striatal cultures. *J. Neurochem.* 2005;95; 715–723.

Fu, HJ; Hu, QS; Lin, ZN; et al. Aluminum-induced apoptosis in cultured cortical neurons and its effect on SAPK/JNK signal transduction pathway. *Brain Res.* 2003;980:11-23.

Garg, G; Adams, JD. Treatment of neuropathic pain with plant medicines. *Chin. J. Integr. Med.* 2012;18:565-570.

Garofalo, K; Penno, A; Schmidt, BP; et al. Oral L-Serine supplementation reduces production of neurotoxic deoxysphingolipids in mice and humans with hereditary sensory autonomic neuropathy type 1. *The J. of Clinical Investigations* 2011;121:4735-45.

Ghribi, O; Dewitt, DA; Forbes, MS; et al. Cyclosporin A inhibits Al-induced cytochrome c release from mitochondria in aged rabbits. *J. Alzheimer's Dis.* 2001;3: 387-91.

Gilardini, A; Avila, RB; Oggioni, N; et al. Myelin structure is unaltered in chemotherapy-induced peripheral neuropathy. *NeuroToxicology* 2012;33:1-7.

Girish, C; Muralidhara. Propensity of *Selaginella delicatula* aqueous extract to offset rotenone-induced oxidative dysfunctions and neurotoxicity in *Drosophila melanogaster NeuroToxicology* 2012;33:444-56.

Gokul, K; Manjunath, MJ; Muralidhara. Exploring the neuroprotective efficacy of *Withania somnifera*: A medicinal plant with diverse biological effects. In: Govil JN (Ed), Recent progress in medicinal plants: Ethnomedicine and therapeutic validation. Studium Press LLC; USA 2012.

Gonthier, B; Allibe, N; Cottet-Rousselle, C; et al. Specific conditions for resveratrol neuroprotection against ethanol induced toxicity. *J. Toxicol;* Article 2012;Id 973134; 11 pages.

Green, DR; Reed, JC. Mitochondria and apoptosis. *Science* 1998;281:1309–1312.

Hacioglu, G; Sevel-Celik, Y; Tanriover, G; et al. Docosahexaenoic acid provides protective mechanism is bilaterally MPTP-lesioned rat model of Parkinson's disease. *Folia Histochemica Et Cytobiologica* 2012;50:228-238.

Halliwell, B. Oxidative stress and neurodegeneration: where are we now? *J. Neurochem.* 2006; 97:1634–58.

Halliwell, B; Gutteridge, JMC. Free Radicals in Biology and Medicine; 4th edn; Oxford University Press; Oxford 2006.

Hemalatha, S; Wahi, AK; Singh, PN; et al. Hypoglycemic activity of *Withania coagulans* Dunal in streptozotocin induced diabetic rats. *J. Ethnopharmacol.* 2004;93(2-3):261-4.

Hidalgo, ME; De la Rosa, C. Antioxidant capacity of eugenol derivatives. *Quim Nova* 2009;32: 1467-1470.

Holocomb, LA; Dhanasekaran, M; Hitt, AR; et al. *Bacopa monniera* extract reduces amyloid levels in PSAPP mice. *J. Alxheimers Dis.* 2006;9:243-51.

Holtzman, D; DeVries, C; Nguyen, H; et al. Maturation of resistance to lead encephalopathy: cellular and subcellular mechanisms. *Neurotoxicology* 1984;5:97-124.

Hosamani, R; Muralidhara. Neuroprotective efficacy of *Bacopa monnieri* against rotenone induced oxidative stress and neurotoxicity in *Drosophila melanogaster*. *Neurotoxicology* 2009;30:977–85.

Hosamani, R; Muralidhara. Prophylactic treatment with *Bacopa monnieri* leaf powder mitigates paraquat induced oxidative perturbations and lethality in *Drosophila melanogaster*. *Indian J. Biochem. Biophys.* 2010;47:75–82.

Houlden, H; King, R; Blake, J; et al. Clinical; pathological and genetic characterization of hereditary sensory and autonomic neuropathy type 1 (HSAN I). *Brain* 2006;129:411–425.

Howes, MJR; Perry, NSL; Houghton PJ. Plants with traditional uses and activities; relevant to the management of Alzheimer's disease and other cognitive disorders. *Phytother Res.* 2003;17:1-18.

Huang, LS; Sun, G; Cobessi, D; et al. 3-Nitropropionic acid is a suicide inhibitor of mitochondrial respiration that; upon oxidation by complex II; forms a covalent adduct with a catalytic base arginine in the active site of the enzyme. *J. Biol. Chem.* 2006;281: 5965–5972.

Huang, SS; Tsai, MC; Chih, CL; et al. Resveratrol reduction of infarct size in Long-Evans rats subjected to focal cerebral ischemia. *Life Sci.* 2001;69:1057-65.

Huang, X; Warren, JT; Buchanan, J; et al. Drosophila Niemann-Pick type C-2 genes control sterol homeostasis and steroid biosynthesis: A model of human neurodegenerative disease. *Development* 2007;134:3733-3742.

Huh, JW; Choi, MM; Lee, JH; et al. Activation of monoamine oxidase isotypes by prolonged intake of aluminium in rat brain. *J. Inorg. Biochem.* 2005;99:2088-91.

Husain, GM; Mishra, D; Singh, PN; et al. Ethnopharmacological review of native traditional medicinal plants for brain disorders. *Pharmacog. Rev.* 2007;1:19-28.

Irish, BP; Khan, ZK; Jain, P; et al. Molecular mechanisms of neurodegenerative diseases induced by human retroviruses: A review. *Am. J. Infect. Dis.* 2009;5:231-58.

Jagatha, B; Mythri, RB; Vali, S; et al. Curcumin treatment alleviates the effects of depletion of glutathione *in vitro* and *in vivo*: Therapeutic implications for Parkinson's disease explained via in silico studies. *Free Radic. Biol. Med.* 2008;44:907-917.

James, SM; Steven, BD. Role of iron in neurotoxicity: a cause for concern in the elderly? *Curr. Opinion in Clinical Nutrition and Metabolic Care* 2009;12:22-29.

Jang, JH; Surh, YJ. Protective effect of resveratrol on beta amyloid induced oxidative PC 12 cell death. *Free Radic. Boil. Med.* 2003;34:1100-1110.

Jimenez-Del-Rio, M; Guzman-Martinez, C; Velez-Pardo, C. The effects of polyphenols on survival and locomotor activity in *Drosophila melanogaster* exposed to iron and paraquat. *Neurochem. Res.* 2010;35:227-238.

Johnson, VJ; Sharma, RP. Aluminum disrupts the proinflammatory cytokine/neurotrophin balance in primary brain rotation-mediated aggregate cultures: possible role in neurodegeneration. *NeuroToxicology* 2003;24:261-8.

Jun-Qing, Y; Bei-Zhong, L; Bai-Cheng, H; et al. Protective effects of meloxicam on aluminum overload-induced cerebral damage in mice. *Eur. J. Pharmacol.* 2006;547:52-8.

Jyoti, A; Sethi, P; Sharma, D; 2007. Bacopa monniera prevents from aluminium neurotoxicity in the cerebral cortex of rat brain. *J. Ethnopharmacol.* 2007;111: 56-62.

Kabuto, H; Tada, M; Kohno, M. Eugenol [2-methoxy-4-(2-propenyl) phenol] prevents 6-hydroxydopamine induced dopamine depression and lipid peroxidation inductivity in mouse straitum. *Biol. Pharm. Bull* 2007;30:423-427.

Kamboj, SS; Sandhir, R. Protective effect of N-acetylcysteine supplementation on mitochondrial oxidative stress and mitochondrial enzymes in cerebral cortex of streptozotocin treated diabetic rats. *Mitochondrion* 2011;11:214-22.

Kandhare, AD; Raygude, KS; Ghosh, P; et al. Therapeutic role of curcumin in prevention of biochemical and behavioural aberration induced by alcoholic neuropathy in laboratory animals. *Neurosci. Letts* 2012; 511:18-22.

Kannarket, G; Lasher, GG; Schiff, D. Neurologic complication of chemotherapy agents. *Current Opinion in Neurology* 2007;20:719-25.

Kaplan, S; Bisleri, G; Morgan, JA; et al. Resveratrol; a natural red wine polyphenol; reduces ischemia-reperfusion induced spinal cord injury. *Ann Thorac. Surg.* 2005;80:2242-49.

Karpinar, DP; Balija, MB; Kugler, S; et al. Prefibrillar alpha-synuclein variants with impaired beta-structure increase neurotoxicity in Parkinson's disease models. *EMBO J.* 2009;28:3256–3268.

Kartnig, T. Clinical application of *Centella asiatica* (L.) Urb. In: Craker LE and simon JE Eds. Herbs; Spices and Medicinal plants. Phoenix: Oxyx Press 1998;145-73.

Kasten-Jolly, Pabello, N; Bolivar, VJ; et al. Developmental lead effects on behaviour and gene expression in male and female BALB/ cAnNTac mice. *NeuroToxicology* 2012;33:1005-10020.

Khan, TM; Benaich, N; Malone, CF; et al. Vincristine and Bortezomib cause outgrowth and behavioral defects in larval zebrafish. *J. Peripheral Nervous system* 2012;17:76-89.

Kirkwood, TB. Understanding the odd science of aging. *Cell* 2005;120:437–447.

Kleijnen, J; Knipschild, P. *Ginkgo biloba* for cerebral insufficiency. *Br. J. Clin. Pharmacol.* 1992;34:352-8.

Koike, H; Iijima, M; Sugiura, M; et al. Alcoholic neuropathy is clinicopathologically distinct from thiamine-deficiency neuropathy. *Ann. Neurol.* 2003;54:19-29.

Kopan, R; Goate, A. Aph-2/Nicastrin: An essential component of gamma-secretase and regulator of Notch signaling and Presenilin localization. *Neuron* 2002;33:321-324.

Krishnan, AV; Goldstein D; Friedlander M; et al. Oxaliplatin induced neurotoxicity and the development of neuropathy. *Muscle nerve* 2005;32:51-60.

Krishnan, AV; Goldstein, D; Friedlander, M; et al. Oxaliplatin and Na+ channel function *in vivo*. *Clin. Cancer Res.* 2006;12: 4481-84.

Kuboyama, T; Tohda, C; Komatsu, K. Neuritic regeneration and synaptic reconstruction induced by withanolide A; *Br. J. Pharmacol.* 2005;144:961–971.

Kuboyama, T; Tohda, C; Komatsu, K. Withanoside IV and its active metabolite sominone attenuate Aâ-induced neurodegeneration. *Eur. J. Neurosci.* 2006;23:1417–1426.

Kulkarni, SK; Verma, AV. GABA receptor mediated anticonvulsant action of *Withania somnifera* root extract; *Indian Drugs* 1993;30:305–312.

Kulkarni, SK; Sharma, A; Verma, A; et al. GABA receptor mediated anticonvulsant action of Withania somnifera root extract; *Indian Drugs* 1993;30:305–312.

Kumar, A; Meena, S; Kalonia, H; et al. Effect of nitric oxide in protective effect of melatonin chronic constriction sciatic nerve injury induced neuropathic pain in rats. *Ind. J. Exp. Biol.* 2011;49:664-671.

Kumar, A; Sharma, SS. NF kB inhibitory action of resveratrol: a probable mechanism of neuroprotection in experimental diabetic neuropathy. *Biochem. Biophys. Res. Comm.* 2010;394:360-65.

Kumar, P; Singh, VK; Singh, DK. Kinetics of enzyme inhibition by active molluscicidal agents ferulic acid; umbelliferone; eugenol and limonene in the nervous tissue of snail *Lymnaea acuminate*. *Phytother Res.* 2009;23:172-177.

Kumar, TP; Antony, S; Soman, S; et al. Role of curcumin in the prevention of cholinergic mediated cortical dysfunctions in streptozotocin induced diabetic rats. *Molecular and cellular endocrinology* 2011;331; 1-10.

Kumar, V; Gill, KD. Aluminium neurotoxicity; Neurobehavioural and oxidative aspects. *Arch. Toxicol.*; 2009;83: 965-78.

Kumar, P; Kumar, A. Effects of root extract of *Withania somnifera* in 3-nitropropionic acid-induced cognitive dysfunction and oxidative damage in rats; *Int. J. Health Res.* 2008;1:139.

Kuwahara, T; Koyama, A; Koyama, S; et al. Systematic RNAi screen reveals involvement of endocytic pathway in neuronal dysfunction in alphasynuclein transgenic *C. elegans*. *Hum. Mol. Genet.* 2008;17:2997-3009.

Kwong, TC. Organophosphate Pesticides: Biochemistry and Clinical Toxicology. *Ther. Drug Monit* 2002;24:144-149.

Lannuzel, A; Lledo, PM; Lamghitnia, HO; et al. HIV-1 envelop proteins 120 and gp160 potentiate NMDA induced [Ca2+]i increase; alter [Ca2+]i homeostasis and induce neurotoxicity in human embryonic neurons. *Eur. J. Neurosci.* 1995; 7:2285-93.

Lee, HP; Pancholi, N; Esposito, L; et al. Early induction of oxidative stress in mouse model of Alzheimer disease with reduced mitochondrial superoxide dismutase activity; *PLoS One* 2012; 7: e28033.

Leibson, T; Lifshitz, M. Organophosphate and Carbamate Poisoning: Review of the Current Literature and Summary of Clinical and Laboratory Experience in Southern Israel. *Isr. Med. Assoc. J.* 2008;10: 767-770.

Leinninger, GM; Backus, C; Sastri, AM; et al. Mitochondria in DRG neurons undergo hyperglycemic mediated injury through Bim; Bax and fission protein Drp1. *Neurobiol. Dis.* 2006;23:11-22.

Leinninger, GM; Russell, JW; van Golen, CM; et al. Insulin like growth factor-1 regulates glucose-induced mitochondrial depolarization and apoptosis in human neuroblastoma. *Cell Death Diff.* 2004;11: 885-96.

Lelkes, E; Unsworth, BR; Lelkes, PI. Reactive oxygen species; apoptosis and altered NGF-signalling in PC 12 pheochromocytoma cells cultured in elevated glucose: An in vitro cellular model for diabetic neuropathy. *Neurotoxicity Res.* 2001;3:189-203.

Lemus-Molina, Y; Sanchez-Gomez, MV; Delgado-Hernandez, R; et al. *Mangifera indica* L. extract attenuates glytamate-induced neurotoxicity on rat cortical neurons. *NeuroToxicology* 2009;30:1053-1058.

Li, J; Krajewski, K; Lewis, RA; et al. Loss-of-function phenotype of hereditary neuropathy with liability to pressure palsies. *Muscle Nerve* 2004;29:205–210.

Li, J; Krajewski, K; Shy, ME; et al. Hereditary neuropathy with liability to pressure palsy: the electrophysiology fits the name. *Neurology* 2002;58:1769–1773.

Lim, TG; Lee, BK; Kwon, JY; et al. Acrylamide up-regulates cyclooxygenase-2 expression through the MEK/ ERK signalling pathway in mouse epidermal cells. *Food Chem. Toxicol.* 2011;49:1259–64.

Lima, PD; Leite, DS; Vasconcellos, MC; et al. Genotoxic effects of aluminium chloride in cultured human lymphocytes treated in different phases of cell cycle. *Food Chem. Toxicol.* 2007;45:1154-9.

Limpeanchob, N; Jaipan, S; Rattanakaruna, S; et al. Neuroptratective effect of *Bacopa monnieri* on beta-amyloid-induced cell death in primary cortical culture. *J. Ethnopharmacol.* 2008;120:112-117.

Little, AA; Edwards, JL; Feldman, EL. Diabetic Neuropathies. *Pract. Neurol.* 2007;7:82-92.

LoPachin, RM; Balaban, CD; Ross, JF. Acrylamide axonopathy revisited. *Toxicol. Appl. Pharmacol.* 2003;188:135–53.

LoPachin, RM. The changing view of Acrylamide neurotoxicity. *NeuroToxicology* 2004;25:617–30.

LoPachin, RM; Gavin, T. Acrylamide induced nerve terminal damage: Relevance to neurotoxic and neurodegenerative mechanisms. *J. Agric. Food Chem.* 2008;56:5994-6003.

Ma, K; Xu, L; Zhang, H; Zhang, S; et al. The effect of *Ginkgo biloba* on the rat retinal ganglion cell survival in the optic nerve crush model. *Act Ophthalmolgica* 2010;88: 553-57.

Mahapatra, SK; Chakraborty, SP; Majumdar, S; et al. Eugenol protects nicotine induced superoxide mediated oxidative damage in murine peritoneal macrophages in vitro. *Eur. J. Pharmacol.* 2009;623: 132-140.

Maher, BF; Stough, C; Shelmerdine, A; et al. The acute effect of combined administration of *Ginkgo biloba* and *Bacopa monnieri* on cognitive function in humans. *Hum. Psychopharmacol.*; 2002;17:163-4.

Maiese, K; Chong, ZZ; Hou, J; et al. Oxidative stress: Biomarkers and novel therapeutics. *Experimental Gerantology* 2010;45:217–34.

Manjunath, MJ; Muralidhara. *Withania somnifera* prophylaxis abrogates rotenone induced oxidative impairments and mitochondrial dysfunctions in striatum and cerebellum of mice: Relevance to Parkinson's disease. *Central Nervous System Agents in Medicine Chemistry.* 2013; 13:1-14.

Mannelli, LDC; Bartolini, A; Ghelardini, C. Neuropathy-induced apoptosis: Protective effect of physostigmine. *J. Neurosci. Res.* 2009;87:1871-76.

Mannelli, LDC; Zanardelli, M; Failli, P; et al. Oxaliplatin induced neuropathy: oxidative stress as pathological mechanism. Protective effect of Silibinin. *The J. of Pain* 2012;13:276-84.

Manning-Bog, AB; McCormack, AL; Li, J; et al. The herbicide paraquat causes up-regulation and aggregation of alpha-synuclein in mice: paraquat and alpha-synuclein. *J. Biol. Chem.* 2002;277:1641-1644.

Mansouri, A; Demeilliers, C; Amsellem, S; et al. Acute ethanol administration oxidatively damages and depletes mitochondrial DNA in mouse liver; brain; heart; and skeletal muscles: protective effects of antioxidants. *J. Pharmacol. Exp. Ther.* 2001;298:737–743.

Marambaud, P; Zhao, H; Davies, P. Resveratrol promotes clearance of Alzheimer's disease amyloid beta peptides. *J. Biol. Chem.* 2005;280: 37377-382.

Martin-Morris, LE; White, K. The Drosophila transcript encoded by the beta-amyloid protein precursor-like gene is restricted to the nervous system. *Development* 1990;110:185-195.

Matrone C; Ciotti MT; Mercanti D; et al. NGF and BDNF signalling control amyloidogenic route and Ab production in hippocampal neurons. *PNAS* 2008;105:13139-144.

Mattson, MP; Haughey, NJ; Nath, A. Cell death in HIV dementia. *Cell Death Differ.* 2005;893-904.

Mattson, MP., Son, TG., Camandola, S. Viewpoint: Mechanisms of action and therapeutic potential of neurophometic phytochemicals. *Dose Response* 2007;5:174-186.

McDonough, KH. Antioxidant nutrients and alcohol. *Toxicology* 2003;189:89–97.

Mehri, S; Abnous, K; Mousavi, SH; et al. Neuroprotective effect of crocin on acrylamide induced cytotoxicity in PC 12 cells. *Cell Mol. Neurobiol.* 2011;30:185–91.

Meredith, GE; Sonsalla, PK; Chesselet, MF. Animal models of Parkinson's disease progression. *Acta Neuropathol.* 2008;115:385-398.

Misra, L; Lal, P; Sangwan, RS; et al. Unusually sulfated and oxygenated steroids from *Withania somnifera*. *Phytochemistry* 2005;66: 2702–2707.

Misra, L; Mishra, P; Pandey, A; et al. Withanolides from *Withania somnifera* roots. *Phytochemistry* 2008;69:1000–1004.

Molliver, DC; Wright, DE; Leitner, ML; et al. IB4-binding DRG neurons switch from NGF to GDNF dependence in early post natal life. *Neuron* 1997;19:849-61.

Murakami, Y; Shoji, M; Hanazawa, S; et al. Preventive effect of bis-eugenol; a eugenol ortho dimer; and inflammatory cytokine expression in macrophages. *Biochem. Pharmacol.* 2003;66:1061–1066.

Murakami, Y; Shoji, M; Hirata, A; et al. Dehydrodiisoeugenol; an isoeugenol dimer; inhibits lipopolysaccharide stimulated nuclear factor kappa B activation and cyclooxygenase-2 expression in macrophages. *Arch. Biochem. Biophys.* 2005;434; 326–332.

Muthuraman, A; Jaggi, AS; Singh, N; et al. Ameleiorative effects of amiloride and pralidoxime in chronic constriction injury and vincristine induced painful neuropathy in rats. *Eur. J. Pharmacol.* 2008;587:104-111.

Naik, AK; Tandan, SK; Dudhgaonkar, SP et al. Role of oxidative stress in pathophysiology of peripheral neuropathy and modulation by Nacetyl- L-cysteine in rats. *Eur. J. Pain* 2006;10:573–579.

Nalini, K; Arroor, AR; Karanth, KS; et al. Effect of Centella asiatica leaf aqueous extract on learning and memory and biogenic amine turnover in albino rats. *Fitoterapia* 1992;62:232-7.

Nandi, A; Chandil, D; Lechesal, R; et al. Bifenthrin causes neurite retraction in the absence of cell death: a model for pesticide associated neurodegeneration. *Med. Sci. Monit* 2006;12:BR 169-73.

Narita, M; Miyoshi, K; Narita, M et al. Involvement of microglia in the ethanol-induced neuropathic pain-like state in the rat. *Neurosci. Lett.* 2007;414:21–25.

Needleman, H. Lead poisoning. *Annu. Rev. Med.* 2004;55:209-222.

Negi, G; Kumar, A; Sharma, SS. Melatonin modulates neuroinflammation and oxidative stress in experimental diabetic neuropathy: Effects on NFk B and Nrf 2 cascades. *J. Pineal Res.* 2011;50:124-131.

Nguyen, N; Lee, SB; Lee, YS et al. Neuroprotection by NGF and BDNF against neurotoxin –exerted apoptotic death in neural stem cells are mediated through Trk receptors; activating PI3-kinase and MAPK pathways. *Neurochem. Res.* 2009;34:942-951.

Nixon, RA; Shea, TB. Dynamics of neuronal intermediate filaments: a developmental perspective. *Cell Motil Cytoskeleton 1992*;22:81–91.

Obrosova, IG. Diabetes and peripheral nerve. *Biochimica et Biophyica Acta* 2009;10:931-40.

Ohyagi, M; Sanjo, N; Yokota, T; et al. Hereditary neuropathy with liability of pressure palsy combined with schwannomas of the peroneal and radial nerves. *J. Neurol.* 2012;259:977-79.

Okouchi, M; Okayama, N; Aw, TY. Hyperglycemia potentiates carbonyl stress induced apoptosis in naïve PC 12 cells: Relationship to cellular redox and activator protease factor-1 expression. *Curr. Neurovasc. Res.* 2005;2:375-86.

Ong, WY; Lu, XR; Hu, CY; et al. Distribution of hydroxynonenal-modified proteins in the kainate-lesioned rat hippocampus: evidence that hydroxynonenal formation precedes neuronal cell death. *Free Radic. Biol. Med.* 2000;28;1214–1221.

Ono, K; Hasegawa, K; Naiki, M; et al. Curcumin has potent anti-amyloidogenic effects for Alzheimer's beta-amyloid fibrils in vitro. *Neurosci.* 2004;75:742–50.

Papanikolaou, NC; Hatzidaki, EG; Belivanis, S; et al. Lead toxicity update. A brief review. *Med. Sci. Monit.* 2005;11:RA329-RA336.

Parihar, MS; Chaudhary, M; Shetty, R; et al. Susceptibility of hippocampus and cerebral cortex to oxidative damage in streptozotocin treated mice prevention by extracts of *Withania somnifera* and Aloe vera; *J. Clin. Neurosci.* 2004; 11:397–402.

Park, HR; Kim, MN; Kim, SJ; et al. Acrylamide induces cell death in neural progenitor cells and impairs hippocampal neurogenesis. *Toxicol. Lett.* 2010; 193:86–93.

Patki, G; Lau, Y-S. Melatonin protects against neurobehavioral and mitochondrial deficits in a chronic mouse model of Parkinson's disease. *Pharmacology Biochemistry and Behavior* 2011;99:704-711.

Patrick, L. Lead toxicity; a review of the literature. Part I: Exposure; evaluation and treatment. *Alternative Medicine Review* 2006;11: 2-22.

Pendleton, RG; Parvez, F; Sayed, M; et al. Effects of pharmacological agents upon a transgenic model of Parkinson's disease in *Drosophila melanogaster*. *J. Pharmacol. Exp. Ther.* 2002;300:91-96.

Perrot, R; Berges, R; Bocquet, A; et al. Review of multiple aspects of neurofilament functions; and their possible contribution to neurodegeneration. *Mol. Neurobiol.* 2008;38:27-65.

Phillips, TJC; Cherry, CL; Cox, S; et al. Pharmacological treatment of painful HIV associated sensory neuropathy: A systematic review and meta-analysis of randomised controlled trials. *PLoS ONE* 2010;5:e14433.

Podratz, JL; Staff, NP; Froemel, D; et al. *Drosophila melanogaster*: A new model to study cisplatin induced neurotoxicity. *Neurobiol. Dis.* 2011; 43:330–37.

Pope, S; Land, JM; Heales, SJR. Oxidative stress and mitochondrial dysfunction in neurodegeneration; cardiolipin a critical target? *Biochimica et Biophysica Acta* 2008;1777:794-799.

Prasad, SN; Muralidhara. Evidence of acrylamide induced oxidative stress and neurotoxicity in *Drosophila melanogaster*: Its amelioration with spice active enrichment: relevance to neuropathy. *NeuroToxicology* 2012;33:1254 – 64.

Prasad, SN; Muralidhara. Neuroprotective efficacy of eugenol and isoeugenol in acrylamide induced neuropathy in rats: Behavioral and biochemical evidence. *Neurochem. Res.* 2013;38:330-345.

Prasad, SN; Raghavendra, R; Lokesh, BR; et al. Spice phenolics inhibit human PMNL 5-lipoxygenase. *Prostaglandins Leukot Essent Fat acids* 2004;70: 521 – 528.

Raabe, TD; Deadwyler, G; Varga, JW; et al. Localization of neuregulin isoforms and erbB receptors in myelinating glial cells. *Glia* 2004;45; 197–207.

Rajeswari, A; Sebesan, M. Inhibition of monoamine oxidase-B by polyphenolic compound; curcumin and its metabolite tetrahydrocurcumin; in a model of Parkinson's disease induced by MPTP neurodegeneration in mice. *Inflammopharmacology* 2008;16:96-99.

Ramesh, T; McGown, A; Al Mashshadi, S; et al. The head is "ON" in the neurons: Neuronal stress in a sod1 Zebra fish model of MND affects neuromuscular junction integrity and causes muscle denervation. *J. Neurol. Neurosurg. Psychiatry* 2012;83:e1.

Raygude, K; Kandhare, AD; Ghosh, P; et al. Evaluation of ameliorative effect of quercetin in the experimental model of alcoholic neuropathy in rats. *Inflammopharmacol.* 2012;20:331-341.

Reed, JC. Bcl-2 family proteins. *Oncogene* 1998;17:3225–3236.

Reilly, MM; Murphy, SM; Laura, M. Peripheral Nerve Society Meeting: Charcot-Marie-Tooth disease. *J. Peripheral. nervous system* 2011;16:1-14.

Reiter, RJ; Manchester, LC; Tan, D-X. Neurotoxins: Free radical mechanisms and melatonin protection. *Curr. Neuropharmacol.* 2010;8:194-210.

Rolo, AP; Palmeira, CM. Diabetes and mitochondrial function: Role of hyperglycemia and oxidative stress. *Toxicology and Applied Pharmacology* 2006;212:167-78.

Rui, P; Sheng, Q; Da-xiang, L; et al. Curcumin improves learning and memory ability and its neuroprotective mechanism in mice. *Chin. Med. J.* 2008;121(9):832-39.

Russo, A; Borrelli, F; Campisi, R; et al. Nitric oxide-related toxicity in cultured astrocytes: effect of *Bacopa monnieri*. *Life Sci* 2003a ;73; 1517-26.

Russo, A; Izzo, AA; Borrelli, F; et al. Free radical scavenging capacity and protective effect of Bacopa monnieri L. on DNA damage. *Phytother Res.* 2003b;17; 870-75.

Sango, K; Suzuki, T; Yanagisawa, H; et al. High glucose induced activation of the polyol pathway and changes of gene expression profiles in immortalized adult mouse Schwann cells IMS32. *J. Neurochem.* 2006;98:446-58.

Saraf, MK; Prabhakar, S; Pandhi, P; et al. *Bacopa monniera* ameleiorates amnesic effects of diazepam qualifying behavioural-molecular partitioning. *Neuroscience* 2008;155:476-84.

Savory, J; Herman, MM; Ghribi, O. Mechanisms of aluminium induced neuro-degeneration in animals: Implications for Alzheimer's disease. *J. Alzheimers Dis.* 2006;10:135-44.

Schemmel, KE; Padiyara, RS; D'Souza, JJ. Aldose reductase inhibitors in the treatment of diabetic peripheral neuropathy: A review. *J. Diabetes and its complications* 2010;24:354-60.

Schliebs, R; Liebmann, A; Bhattacharya, SK; et al. Systemic administration of defined extracts from *Withania somnifera* (Indian Ginseng) and Shilajit differentially affects cholinergic but not glutamatergic and GABAergic markers in rat brain; *Neurochem. Int.* 1997;30: 181– 190.

Selvaraj, S; Watt, JA; Singh, BB. TRPC1 inhibits apoptotic cell degeneration induced by dopaminergic neurotoxin MPTP/ MPP(+). *Cell Calcium.* 2009;46:209–218.

Sharifi, AM; Mousavi, SH; Farhadi, M; et al. Study of high glucose-induced apoptosis in PC12 cells: role of bax protein. *J. Pharmacol. Sci.* 2007;104:258-62.

Sharma, S; Kulkarni, SK; Agrewala, JN; et al. Curcumin attenuates thermal hyperalgesia in a diabetic mouse model of neuropathic pain. *Euro. J. of Pharmacol.* 2006;536:256-261.

Sherer, TB; Betarbet, R; Testa, CM; et al. Mechanism of toxicity in rotenone models of Parkison's disease. *J. Neurosci.* 2003a;23:10756–10764.

Sherer, TB; Kim, JH; Betarbet, R; et al. Subcutaneous rotenone exposure causes highly selective dopaminergic degeneration and alpha-synuclein aggregation. *Exp. Neurol.* 2003b:179; 9-16.

Shi, J; Ma, Y; Zheng, M; et al. Effect of sub-acute exposure to acrylamide on GABAergic neurons and astrocytes in weaning rat cerebellum. *Toxicol. Ind. Health* 2011; 1-11.

Shinomol, GK; Bharath, MMS; Muralidhara. Neuromodulatory propensity of *Bacopa monnieri* leaf extract against 3-nitropropionic acid-induced oxidative stress: *In vitro* and *in vivo* evidences. *Neurotox Res.* 2012;22:102-114.

Shinomol, GK; Hosamani, R; Muralidhara. Prophylaxis with *Centella asiatica* confers protection to prepubertal mice against 3-nitropropionic acid induced oxidative stress in brain. *Phytpther Res.* 2010;24:885-92.

Singh, RH; Singh, L. Studies of the anti-anxiety effect of the medhyarasayana drug Brahmi (*Bacopa monnieri*) *Res. Ayur Sid.* 1980;1:133-48.

Smith, HS. Treatment considerations in painful HIV-related neuropathy. *Pain Physician* 2011;14:E505-E524.

Sofroniew, MV; Howe, CL; Mobley, WC. Nerve growth factor signaling; neuroprotection; and neural repair. *Annu. Rev. Neurosci.* 2001;24:1217–1281.

Somayajulu-Nitu, M; Sandhu, JK; Cohen, J; et al. Paraquat induces oxidative stress; neuronal loss in substania nigra region and Parkinsonism in rats: Neuroprotection and amelioration of symptoms by water soluble formulation of Coenzyme Q10. *BMC Neuroscience* 2009;10:88.

Srinivasan, K. Spices for taste and flavour: Nutraceuticals for human health. In: Amit Krishna De; editor. Spices: the elixir of life. New Delhi: Original Publications 2011; 43-62.

Stadler, RH; Blank, I; Vorga, N; et al. Acrylamide from reaction products. *Nature* 2002;419:449–50.

Subartha, M; Shila, S; Anusuya Devi, M; et al. Emerging role of *Centella asiatica* in improving age-related neurological antioxidant status. *Exp. Gerontol.* 2005;40:707-15.

Symth, K; Affandi, JS; McArthur, JC; et al. Prevalence of and risk factors for HIV associated neuropathy in Melbourne; Australia 1993-2006. *HIV Med.* 2007;367-73.

Szigeti, K; Lupski, JR. Charcot-Marie-Tooth disease. *Eur. J. Human Genetics* 2009;17:703-710.

Tardiff, RG; Gargas, ML; Kirman, CR; et al. Estimation of safe dietary intake levels of acrylamide for humans. *Food Chem. Toxicol.* 2010;48:658–67.

Tareke, E; Lyn-Cook, BD; Duhart, H; et al. Acrylamide decreased dopamine levels and increased 3-nitrotyrosine (3-NT) levels in PC 12 cells. *Neurosci. Lett.* 2009;458:89–92.

Tareke, E; Rydberg, P; Karlsson, P; et al. Analysis of acrylamide; a carcinogen formed in heated foodstuffs. *J. Agric. Food Chem.* 2002;50:4998–5006.

Tiwari, V; Kuhad, A; Chopra, K. Neuroprotective effect of vitamin E isoforms against chronic alcohol induced peripheral neurotoxicity; Possible involvement of oxidative-nitrodative stress. *Phytother Res.* 2012;26:1738-45.

Vettori, A; Bergamin, G; Moro E; et al. Developmental defects and neuromuscular alterations due to mitofusin 2 gene (MFN2) silencing in zebrafish: a new model of Charcot-Marie-Tooth type 2A neuropathy. *Neuromuscular disorders* 2011;21:58-67.

Viala, A; Cano, JP; Duranid, A; et al. Study in animals of transcutaneous penetration of labelled active principles in the titrated extract of *Centella asiatica* applied on the skin in two topical forms. *Therapie*1977;32:573-84.

Vincent, AM; Brownlee, M; Russell, JW. Oxidative stress and programmed cell death in diabetic neuropathy. *Ann. NY Acad. Sci.* 2002;959:368-83.

Vincent, AM; Feldman, EL. Can drug screening lead to candidate therapies for testing in diabetic neuropathy. *Antioxidants and Redox signalling* 2008;10:387-93.

Vincent, AM; McLean, LL; Backus, C; et al. Short term hyperglycemia produces oxidative damage and apoptosis in neurons. *FASEB J.* 2005;19:638-40.

Vincent, AM; Olzmann, JA; Brownlee, M; et al. Uncoupling proteins prevent glucose-induced neuronal oxidative stress and programmed cell death. *Diabetes* 2004b; 53:726-34.

Vincent, AM; Russell, JW; Low, P; et al. Oxidative stress in the pathogenesis of diabetic neuropathy. *Endocrine Reviews* 2004a;25:612-28.

Wang, L; Xi, G; Keep, RF; et al. Iron enhances the neurotoxicity of amyloid b. *Translational Stroke Research* 2012;3:107-113.

Wattanathorn, J; Mator, L; Muchimapura, S; et al. Positive modulation of cognition and mood in the healthy elderly bolunteer following the administration of Centella asiatica. *J. Ethnopharmacol.* 2008;116: 325-32.

Wie, MB. Eugenol inhibits excitotoxins induced delayed neurotoxicity; oxidative damage and convulsion. *J. Toxicology and Public Health* 2006;22:275-82.

Wie, MB; Won, MH; Lee, KH; et al. Eugenol neuronal cells from excitotoxic and oxidative injury in primary cortical cultures. *Neurosci. Lett.* 1997;225:93-96.

Wild, S; Roglic, G; Green, A; et al. Global prevalence of diabetes: estimates for the year 2000 and projections for 2030. *Diabetes Care* 2004;27:1047-53.

Wimpissinger, B; Berisha, F; Garhoefer, G; et al. Influence of *Ginkgo biloba* on ocular blood flow. *Acta Ophthalmol. Scand.* 2007;85:445-9.

Windebank, AJ. *Current Opinion in Neurology* 1999;12:565-71.

Wolf, CI; Mannion, RJ. Neuropathic pain; aetiology; symptoms; mechanisms and management. *Lancet* 1999;353:1959-64.

Won, MH; Lee, JC; Kim, YH; et al. Postischemic hypothermia induced by eugenol protects hippocampal neurons from global ischemia in gerbils. *Neurosci. Lett.* 1998;254:101-104.

Xiao, WH; Bennett, GJ. Effects of mitochondrial poisons on the neuropathic pain produced by the chemotherapeutic agents; paclitaxel and oxaliplatin. *Pain* 2012;153:704-709.

Xu, Y; Qui, H-Q; Liu, H; et al. Effects of Koumine; an alkaloid of *Gelsmium elegans* Benth.; on *inflammatory and neuropathic pain models and possible mechanisms with allopregnanolone*. 2012;101:504-514.

Yeh, JL; Hsu, JH; Hong, YS; et al. Eugenolol and glyceryl-isoeugenol suppress LPS-induced iNOS expression by down-regulating NF-kappaB AND AP-1 through inhibition of MAPKS and AKT/IkappaBalpha signaling pathways in macrophages. *International Journal of Immunopathology and Pharmacology* 2011;24:345-356.

Yener, Y; Kalipci, E. The carcinogenic effects of acrylamide formed during cooking of some foods. *Academic J. Cancer Res.* 2009;2:25–32.

Yerdelen, D; Koc, F; Uysal, H. Strength-duration properties of sensory and motor axons in alcoholic polyneuropathy. *Neurol. Res.* 2008;30:746-750.

Yeste-Valasco, M; Alvira, D; Sureda, FX; et al. DNA low density array analysis of colchicine neurotoxicity in rat cerebellar granular cells. *NeuroToxicology* 2008;29:309-317.

Yousef, MI; El-Demerdash, FM. Acrylamide induced oxidative stress and biochemical perturbations in rats. *Toxicology* 2006;219:133–41.

Yousuf, S; Atif, F; Ahmad, M; et al. Resveratrol exerts its neuroprotective effect by modulating mitochondrial dysfunctions and associated cell death during cerebral ischemia. *Brain Res.* 2009;23:242-253.

Yu, S; Son, F; Yu, J; et al. Acrylamide alters cytoskeletal protein in sciatic nerve. *Neurochem. Res.* 2006;31:1197–204.

Zecca, L; Youdim, MB; Riederer, P; et al. Iron; brain ageing and neurodegenerative disorders. *Nat. Rev. Neurosci.* 2004;5:863–873.

Zelenka, M; Schafers, M; Sommer, C. Intraneural injection of interleukin-1beta and tumor necrosis factor-alpha into rat sciatic nerve at physiological doses induces signs of neuropathic pain. *Pain* 2005;116:257–263.

Zhang, L; Gavin, T; Barber, DS; et al. Role of the Nrf2-ARE pathway in acrylamide neurotoxicity. *Toxicol. Lett.* 2012;205:1-7.

Zheng, H; Xiao, WH; Bennett, GJ. Mitotoxicity and bortezomib-induced painful peripheral neuropathy. *Exp. Neurol.* 2012;238:225-234.

Zheng, H; Xiao, WH; Bennett GJ. Functional deficits in peripheral nerve mitochondria in rats with paclitaxel- and oxaliplatin- evoked painful peripheral neuropathy. *Exp. Neurol.* 2011a;232:154-61.

Zheng, M; Gorelenkova, O; Yang, J; et al. A Liquid Phase Based *C. elegans* Behavioral Analysis System Identifies Motor Activity Loss in a Nematode Parkinson's Disease Model. *J. Neurosci. Methods* 2011b;204: 234-237.

Zhong, J; Yao, W; Lee, W. Cesium chloride protects cerebellar granule neurons from apoptosis induced by low potassium. *Int. J. Devl Neurosci.* 2007;25:359-365.

Zhu, YJ; Zeng, T; Zhu, Y-B; et al. Effects of acrylamide on the nervous tissue antioxidant system and sciatic nerve electrophysiology in the rat. *Neurochem. Res.* 2008;33:2310–17.

Zimmermann, M. Pathobiology of neuropathic pain. *Eur. J. Pharmacol.* 2001;429:23–37.

Zochodne, DW; Ramji, N; Toth, C. Neuronal targeting in diabetes mellitus: A story of sensory neurons and motor neurons. *Neuroscientist* 2008;14:311-18.

In: Advances in Neuropathy Research
Editors: M. O. Brown and B. A. Budd
ISBN: 978-1-62618-400-8

Chapter 2

Cardiac Autonomic Dysfunction in Type 2 Diabetes Mellitus: Insights from One Minute Electrocardiogram Studies

***N. K. Subbalakshmi*[1,*] *and Prabha Adhikari*[2]**

[1]Department of Physiology, Kasturba Medical College Mangalore, Manipal University, Mangalore, Karnataka State, India

[2]Department of Medicine, Kasturba Medical College Mangalore, Manipal University, Mangalore, Karnataka State, India

Abstract

Diabetes mellitus is a metabolic disorder affecting various organs of the body. In many patients with diabetes mellitus, typical symptoms associated with the disease manifest only after sufficient cumulative effect of diabetes mellitus. As a result, diabetes-related complications and associated clinical conditions may be present by the time it is clinically diagnosed. One such diabetic complication which remains subclinical is autonomic neuropathy. Diabetic neuropathy is associated with morbidity

* Corresponding author: N. K. Subbalakshmi. Telephone Number: (91) 9483910879; Fax: (91) 824 242 8183; E-mail: rao.subbalakshmink@rediffmail.com; Postal Address: Dr. Subbalakshmi NK, Department of Physiology, Kasturba Medical College, PO Box 53, Light House Hill Road, Hampanakatta, Mangalore 575001, Karnataka, India.

and mortality. In the light of increased mortality rate associated with diabetic autonomic neuropathy, and its possible involvement in other complications it is of utmost importance to assess cardiac autonomic function periodically for long-term treatment strategies. Among the different non-invasive techniques for assessing autonomic status, heart rate variability has emerged as a major tool in the diagnostic and monitoring armamentarium.

Currently, based on the recommendations of task force of the European Society of Cardiology and the North American Society of Pacing and Electrophysiology time domain measures of heart rate variability are typically obtained from a routine 24-h ambulatory electrocardiogram. However heart rate variability parameters derived from much lesser duration may be more appropriate for routine clinical assessment of cardiac autonomic dysfunction. This article portrays insights from one minute electrocardiogram studies on assessment of cardiac autonomic dysfunction in type 2 diabetes mellitus; wherein, heart rate response and Inspiratory : Expiratory ratio are derived from lead II electrocardiogram while deep breathing at 6 respiratory cycles per minute; rMSSD and SDNN are derived from lead II electrocardiogram while breathing spontaneously in resting supine position.

Analysis of heart rate response to deep breathing at 6 respiratory cycles per minute at baseline reveals that, among the clinical characteristics of type 2 diabetes mellitus, somatic neuropathy is an independent predictor of cardiac parasympathetic dysfunction. Age and female gender are the other predictors of reduced heart rate response to deep breathing in type 2 diabetes mellitus.

Monitoring cardiac autonomic function over one year period reveals that cardiac autonomic function deteriorates with time and essentially in presence of somatic neuropathy and cardiovascular disease. rMSSD an indicator of parasympathetic activity detects deterioration in cardiac autonomic dysfunction coexisting with somatic neuropathy and cardiovascular disease. But Expiratory: Inspiratory ratio another indicator of parasympathetic function is a suitable marker in monitoring cardiac autonomic dysfunction coexisting with somatic neuropathy. Similarly, SDNN an indicator of parasympathetic and sympathetic selectively quantifies cardiac autonomic dysfunction coexisting mainly with cardiovascular disease. Thus these findings suggest the possible utility of time domain measures of heart rate variability derived from one minute electrocardiogram in routine clinical assessment of cardiac parasympathetic dysfunction in type 2 diabetes mellitus.

Keywords: Cardiac parasympathetic dysfunction, HRV with deep breathing, E: I ratio, rMSSD, SDNN

Introduction

The diabetic neuropathies are heterogeneous, affecting different parts of the nervous system that present with diverse clinical manifestations. They may be focal or diffuse. Most common among the neuropathies are chronic sensorimotor distal symmetric polyneuropathy and the autonomic neuropathies [1]. One of the most serious complications of diabetes is cardiovascular autonomic neuropathy [2] which encompasses damage to the autonomic nerve fibers that innervate the heart and blood vessels, resulting in abnormalities in heart rate control and vascular dynamics [3]. Impaired autonomic regulation of cardiovascular functioning is an independent predictor of diabetes-related mortality [4]. The prevalence of cardiac autonomic neuropathy (CAN) is approximately 25% and 34% among individuals with type 1 and type 2 diabetes respectively; when diagnostic criteria are based upon at least two abnormalities in autonomic function tests [5]. Cardiovascular autonomic neuropathy has been linked to postural hypotension, exercise intolerance, enhanced intraoperative cardiovascular lability, increased incidence of asymptomatic ischemia, myocardial infarction, and decreased likelihood of survival after myocardial infarction [6]. Among the different non-invasive techniques for assessing cardiac autonomic status, heart rate variability (HRV) has emerged as a major tool in the diagnostic and monitoring armamentarium. This chapter will deal with: 1. a brief over view of different methods of HRV analysis. 2. Discuss the determinants of heart rate response to deep breathing (HRV with deep breathing) and corrected Q-T interval (QTc) among the clinical characteristics of type 2 diabetics. 3. Reflect on utility of certain time domain measures of HRV and resting heart rate in monitoring type 2 diabetics. 4. A brief over view of treatment of cardiac autonomic neuropathy.

Measurement of Heart Rate Variability (HRV)

There are two approaches to measurement of HRV: analysis in the frequency or in the time domain. The frequency domain analysis addresses the question "what are the underlying rhythms?" Frequency domain analysis is used to partition the total variance of the heart rate into the variance accounted for by underlying groups of frequencies [7]. However, to carry out spectral analysis correctly, a noise-free signal is required. Artifacts and extra-systoles must therefore be corrected. Since this correction is associated with a loss of data, the variance of the RR interval is underestimated in each case [8]. Another disadvantage is that, due to the different technical properties of each HRV-measuring device in the field of spectral analysis, specific reference values have to be collected [9]. Generally, the spectrum determined is divided into four frequency ranges, which enable an assessment of the sympathetic and parasympathetic influences on heart rate [10, 11]. Clinical observations and experiments have proven that high-frequency components (HF; 0.15–0.4 Hz) are predominantly traced back to vagal influences [12, 13]. The lower-frequency component (LF; 0.04–0.15 Hz) is considered as being sympathetically modulated by some authors [13-15]. However, numerous publications have proven that both sympathetic and parasympathetic influences play a part [16-20]. Time domain measures of HRV are of two types. Simple time domain measures and statistical methods. Simple time domain variables that can be calculated include the mean NN interval, the mean heart rate, the difference between the longest and shortest NN interval, the difference between night and day heart rate. Other time domain measurements that can be used are variations in instantaneous heart rate secondary to respiration, tilt and Valsalva manoeuvre. Statistical methods calculate HRV from a series of instantaneous heart rates or cycle intervals, particularly those recorded over longer periods. Statistical methods may be divided into two classes, (a) those derived from direct measurements of the NN intervals or instantaneous heart rate such as SDNN and (b) those derived from the differences between NN intervals such as rMSSD [11]. Interestingly every time domain indices of HRV has at least one frequency domain index of HRV that correlates with it sufficiently to be used as a surrogate [21]. r-MSSD correlate with HF power, SDNN index correlate significantly with total power [22]. In experiments comparing spectral analysis and standardized autonomic function tests, there was a high correlation between the results of both techniques [9]. The task force on HRV recommends that, in order to standardize clinical studies, only two types of HRV measurements should be used (1) short term measurements on 5-minute electrocardiograms made under physiologically stable conditions, processed by frequency-domain methods, and (2) 24-hour recordings processed by time-domain methods [11]. However, certain population based studies have reported an association between decreased HRV measured in 20 second electrocardiograms [23] and 10 seconds electrocardiograms and cardiac mortality [24]. These observations suggest that calculation of time domain measures of HRV from electrocardiogram tracing of shorter duration may also be of value in evaluating and monitoring cardiac autonomic dysfunction.

Determinants of Cardiac Parasympathetic Dysfunction among the Clinical Characteristics of Type 2 Diabetics

Diabetes mellitus is a metabolic disorder affecting various organs of the body. In many patients with diabetes mellitus, typical symptoms associated with the disease manifest only after sufficient cumulative effect of diabetes mellitus. As a result, diabetes-related complications and associated clinical conditions may be present by the time it is clinically diagnosed. One such diabetic complication which remains subclinical is autonomic neuropathy. In the light of increased mortality rate associated with diabetic autonomic neuropathy, and its possible involvement in other complications [25-27], it is crucial to detect early changes in the autonomic tone, as well as its relationship to other complications, for long-term treatment strategies. It is generally believed that the parasympathetic dysfunction manifests much earlier, before the involvement of the sympathetic fibers [28]. In the intact heart, parasympathetic fibers are inhibitory and sympathetic fibers are excitatory. Inhibitory actions of cardiac parasympathetic nerves are reported to provide electrical stability to the heart, thus preventing ventricular tachycardia in humans [28]. Vagal nerve traffic cannot be measured directly in humans. The assessment of HRV has thus become widely used indirect measure of cardiac vagal nerve function. Wheeler and Watkins drew attention to the important role of vagal cardiac innervation in the mediation of HRV. They documented the reduction or loss of HRV of diabetics with autonomic neuropathy and demonstrated that HRV was abolished by atropine but unaltered by sympathetic blockade. These authors hypothesized that the loss of HRV associated with diabetic neuropathy was due to vagal cardiac denervation. This important observation drew clinical attention to the important role played by the vagus nerve in the mediation of HRV [29]. HRV while deep breathing is the most commonly used test and is part of most autonomic screening batteries; it is currently the closest to an optimal cardiovagal measure available [30, 31]. It has been examined in diverse conditions and is a sensitive marker for diabetic autonomic neuropathy [32].

HRV with Deep Breathing and Age Related Normal Range

The effect of ageing on HRV with deep breathing is well known. Studies involving to date have found progressive reduction in the response with increasing age [33-35]. The tests of Ewing battery, which were developed with their normal ranges before this was appreciated, contained only one normal range regardless of age [25]. Thus they stand to generate false negative results among younger patients and false positive results among the older age group. Normal values reported by later workers for each age range of heart rate change are: 10–40 years: > 18 beats/minute; 41–50 years: >16 beats/minute; 51–60 years: > 12 beats/minute; and 61–70 years: > 8 beats/minute [30]

Gender Related Differences in HRV with Deep Breathing

Gender related differences in HRV with deep breathing have been studied in type 1 and type 2 diabetes mellitus. With regard to whether either gender is more likely to develop autonomic dysfunction in type 1 diabetes mellitus, the literature has revealed conflicting reports. In the Diabetes Control and Complication Trial (DCCT), the presence of autonomic neuropathy correlated with male gender, along with age and duration of diabetes mellitus [36]. Jaffe et al. showed that the male gender is predictive of depressed HRV, in addition to age, duration of diabetes mellitus, and retinopathy [37]. However, in another study in type 1 diabetic individuals, the female gender was found to be an independent determinant of autonomic dysfunction [38]. But in type 2 diabetes mellitus, Toyry et al. have observed diminished parasympathetic function in female patients [39]. In a recent study by subbalakshmi et al., female gender was one of the independent predictors of reduced HRV with deep breathing [40]. Thus it appears that female type 2 diabetics are at greater risk of cardiac autonomic dysfunction.

Association between Duration of Diabetes and HRV with Deep Breathing

Association between duration of diabetes mellitus and reduced HRV is unclear. On one hand, Dyberg et al. have observed increased frequency of diabetic neuropathy with longer duration of diabetes mellitus [41]. Similarly, Gundersen and Neubauer [42] and O'Brien et al. [34] have found an association between the duration of diabetes mellitus and reduced HRV in patients with type 1 diabetes mellitus. On the other hand, in Straub et al.'s study involving patients with type 1 and type 2 diabetes mellitus, the duration of diabetes mellitus did not correlate with the results of cardiovascular tests [43]. Likewise, Veglio et al. did not find any correlation between the duration of diabetes mellitus and HRV in type 2 diabetic patients [44]. Furthermore, Lehtinen et al. detected an impaired HRV in patients with newly-diagnosed type 2 diabetes mellitus [45]. Pfeifer et al. detected diminished autonomic dysfunction in patients with type 1 and type 2 diabetes mellitus shortly after the diagnosis of diabetes mellitus [46]. A recent study by certain researchers did not find independent association between duration of diabetes mellitus and cardiac autonomic dysfunction in type 2 diabetes mellitus. But they too have observed negative correlation between duration of diabetes mellitus and HRV with deep breathing [40]. In their study, type 2 diabetics were unselected with respect to age and diabetic complications. The proportion of diabetics with peripheral neuropathy and age of diabetics was higher with increasing duration of diabetes. Therefore they attribute the observed negative correlation between the duration of diabetes mellitus and HRV to the negative impact of advancing age and presence of diabetic complications, in particular somatic neuropathy on HRV [40].

Influence of Cardiovascular Disease (CVD) on HRV with Deep Breathing

CVD is one of the major complications of diabetes mellitus. People with diabetes die of CVD two hold higher compared to patients free from diabetes mellitus. HRV analysis in patients with diabetes mellitus reveals that hypertension is an independent predictor of cardiac autonomic dysfunction in type 1 diabetics [38]. In the study of Subbalakshmi et al. negative correlation is observed between hypertension and HRV in type 2 diabetes mellitus. However, in this study, independent association between hypertension and reduced HRV could not be established [40]. Similarly they have observed negative correlation between coronary artery disease and cardiac parasympathetic dysfunction in type 2 diabetics; But did not find independent association between coronary artery disease and reduced HRV [40]. In type 2 diabetics CVD often coexist with somatic neuropathy. Secondly frequency of CVD including hypertension increases with ageing [40]. Therefore, they reason out that the reduced HRV observed in diabetic patients with coronary artery disease and hypertension may be to certain extent attributed to the influence of advancing age of the patients and presence of somatic neuropathy.

Influence of Diabetic Retinopathy on HRV with Deep Breathing

Diabetic retinopathy is an independent predictor of cardiac autonomic dysfunction in type 1 diabetes mellitus [47] but not in patients with type 2 diabetes mellitus [40]. However negative correlation has been observed between diabetic retinopathy and HRV in type 2 diabetics [40, 48, 49].

Association between Somatic Neuropathy and HRV with Deep Breathing

Somatic neuropathy is an independent predictor of cardiac parasympathetic dysfunction in type 2 diabetes mellitus [40]. HRV with deep breathing declines with severity of somatic neuropathy [40]. At the same time, HRV may be reduced in diabetic patients free of all complications compared to non-diabetic healthy subjects [40]. Thus in some patients with diabetes mellitus cardiac parasympathetic dysfunction may either be isolated or may precede other complications.

Influencing Factors of Corrected QT Interval (QT_c) among the Clinical Characteristics of Type 2 Diabetics

Relevance of Measurement of Q-T Interval in Diabetes Mellitus and Diabetic Neuropathy

Cardiovascular disease (CVD) accounts for the majority of deaths for people with type 2 diabetes mellitus. CVD is a broad term which includes any condition causing pathological changes in blood vessels, cardiac muscle or valves and cardiac rhythm. The electrocardiogram (ECG) offers a quick, non-invasive clinical and research screen for the early detection of CVD [50]. Prolonged QT and QTc intervals are considered as reliable predictors of heart disease and fatal ventricular arrhythmias [51-53]. Prevalence of prolonged QT interval is higher in people with type 1 and type 2 diabetes as compared to non-diabetics [54-56]. The prevalence of QT prolongation has been reported to be as high as 16% in type1 and 26% in type 2 diabetes [55]. Prolonged QTc is reported to be an independent marker for coronary heart disease in type 1 and type 2 diabetes and have been demonstrated to be highly significant predictor of cardiac death even in newly diagnosed type 2 diabetes [57]. A positive linear relationship exists between QTc interval prolongation and diabetic cardiac autonomic neuropathy severity in diabetic population [55].

Influence of Gender on QTc

Healthy Non-diabetic and diabetic females have longer Q-Tc compared to non-diabetic and diabetic males [58]. Thus studying the influence of clinical characteristics of type 2 diabetes on QTc would be more appropriate, if these influences are investigated in male and female diabetics separately. Realizing the importance of sex difference in defining the upper limit of normal value for QTc, Moss AJ has suggested 3-level classification of QTc adjusted for age and gender [59].

Referral value of this classification is: for men: - normal: < 0.43, borderline: 0.43-0.45, abnormal: > 0.45; for women: - normal: < 0.45, borderline: 0.45-0.47, abnormal: > 0.47.

More recently, Subblakshmi et al. have examined and confirmed the validity of these referral values in non-diabetic healthy subjects. Based on these referral values they have observed 11.56 % of type 2 diabetics with prolonged QTc [58].

Influence of Hypertension and Retinopathy on QTc in Type 2 Diabetics

Proportion of type 2 diabetics with prolonged QTc is mainly higher in diabetics with somatic neuropathy [58]. Nevertheless, QTc of either male or female diabetics with hypertension and diabetic retinopathy is higher compared to non-diabetic healthy subjects [58].

Thus it seems that, although hypertension and retinopathy did not induce abnormal QTc in diabetics, these determinants enhance a greater possibility of abnormal QTc.

Influence of Cardiac Autonomic Dysfunction on QTc in Male and Female Type 2 Diabetics

Several studies have found direct association between the extent of cardiac autonomic neuropathy and Q-T interval prolongation [60-63]. Subbalakshmi et al. have found negative correlation between QTc and HRV with deep breathing in male but not in female type 2 diabetics [58]. Veglio et al. have found association between QTc prolongation and impaired HRV with deep breathing in male but not in female type 1diabetics [56].

Measures of HRV with deep breathing are the most sensitive indicator of parasympathetic function [64]. Cardiac parasympathetic nerves are reported to provide electrical stability to the heart and thus preventing ventricular tachycardia in humans [28]. Thus the decline in vagal activity could be one of the factors in prolonging QTc interval and predisposing to ventricular tachycardia in men.

Influence of Somatic Neuropathy on QTc in Male and Female Type 2 Diabetics

QTc of female type 2 diabetics with somatic neuropathy is reported to be higher compared to non-diabetic healthy subjects and diabetic subjects free from somatic neuropathy [58]. In this study QTc prolongation was positively correlating with severity of somatic neuropathy in male and female type 2 diabetics. In their study population, 85.71% of male diabetics with impaired Q-Tc were having somatic neuropathy; However, QTc of male type 2 diabetics did not differ from non-diabetic healthy subjects and diabetics free from somatic neuropathy [58]. Thus it could be said that QTc prolongation is associated with somatic neuropathy but with certain sex difference in its clinical manifestation [58].

Influence of Ischemic Heart Disease on QTc in Type 2 Diabetics

In the Eurodiab IDDM complication study QTc prolongation was found to be associated with ischemic heart disease and this association was stronger in males than in females [56]. Whereas T.C. Ko Gary et al. found association between prolonged QTc and ischemic heart disease in Chinese type 2 diabetic women [65]. However another study in type 2 diabetes mellitus did not find sufficient evidence in support of association between QTc prolongation and ischemic heart disease [58].

Influence of Insulin on QTc

Certain researchers probing into the influence of insulin on QTc have observed that physiological hyperinsulinemia acutely prolongs ventricular repolarization independent of insulin sensitivity. Their study finding is that both insulin-induced hypokalemia and adrenergic activation contribute to this effect [66]. In the study of Subbalakshmi et al. too, type 2 diabetic subjects receiving insulin therapy had longer QTc compared to type 2 diabetics not receiving insulin therapy [58]. However, they are of the opinion that, further randomized drug trials are required to establish the association between QTc prolongation and therapeutic doses of insulin in type 2 diabetes mellitus.

Influence of Ageing on QTc

Ageing has profound effect on autonomic nervous system. Accordingly, all the studies involving to date have found progressive reduction in HRV with deep breathing with increasing age in healthy subjects and in patients with diabetes mellitus [30, 32, 34]. Several studies have also pointed out that QTc prolongation is a feature of autonomic dysfunction [67-69]. But a study by Subbalakshmi et al. has not found any association between QTc and age, QTc and HRV with deep breathing in non-diabetic healthy subjects of either sex [58]. Nevertheless, in their study, QTc was positively correlated with age and negatively correlated with HRV with deep breathing in diabetic men. Thus they are of the opinion that diabetes mellitus related cardiac autonomic dysfunction may contribute to the QTc prolongation and not the physiological effect of ageing on autonomic nervous system [58].

Progression of Diabetic Cardiac Autonomic Neuropathy

Several studies in the past have investigated the progression of diabetic autonomic neuropathy. Most of these studies have documented decline in measures of HRV over a period of time. Karmitsos DT et al. have observed decline in HRV in one year period. In their study patients with better glycemic control exhibited deterioration of diabetic autonomic neuropathy at the same time period as those with poor glycemic control [70]. In Mackey JD et al.'s study there was a small deterioration of autonomic function, with a decrease in HRV score of 1.0 per year in IDDM patients [71].

In the study of Sampson MJ et al. HRV declined 1.02 ± 0.47 per annum with an initially normal HRV in IDDM patients [72]. Dyck et al. have reported 2.05 beats per minute in IDDM and 1.56 beats per minute in NIDDM patients [73].

Progression of Cardiac Autonomic Dysfunction in Relation to Presence of Somatic Neuropathy and Cardiovascular Disease

More recently subbalakshmi et al. have studied the cardiac autonomic dysfunction in relation to presence of somatic neuropathy and cardiovascular disease over a period of one year.

In their study, in type 2 diabetics with somatic neuropathy E: I ratio and r-MSSD declined significantly. SDNN did not change appreciably [74]. E: I ratio and r-MSSD quantify parasympathetic mediated cardiac activity [75-77]. SDNN is influenced both by sympathetic and parasympathetic activity [78].

Thus they suggest that, in diabetics with somatic neuropathy, predominantly cardiac parasympathetic function deteriorates in one year period [74]. In their study, type 2 diabetics with cardiovascular disease alone, r-MSSD and SDNN declined significantly in one year period. But E: I ratio did not change appreciably [74]. r-MSSD reveals alterations in autonomic tone that are predominantly vagally mediated and regarded as a reliable index of cardiac parasympathetic activity [76, 77].

However the other well-known measure of vagal activity, namely E: I ratio did not change appreciably in their type 2 diabetics with cardiovascular disease. Thus they suggest that r-MSSD could be of value in monitoring subtle changes in cardiac parasympathetic function which fails to reflect in E: I ratio [74]. The standard deviation of the mean R-R interval replicates all the cyclic components responsible for variability in the period of recording and is regarded to reflect both the sympathetic and the parasympathetic influence on HRV [78]. At the same time cardiac sympathetic over activity is reported to be associated with CVD [79, 80].

Thus Subbalakshmi et al. speculate that cardiovascular complication in type 2 diabetes mellitus may involve dysfunction of sympathetic nerves in addition to parasympathetic nerve dysfunction.

Resting Heart Rate and Progression of Cardiac Autonomic Dysfunction

Resting heart rate is higher in patients with type 2 diabetes mellitus compared to non-diabetic healthy subjects. Monitoring resting heart rate over one year period in patients with type 2 diabetes mellitus have revealed that heart rate rises appreciably in subjects either with cardiovascular or somatic neuropathy complication; but in presence of combined cardiovascular and somatic neuropathy complications resting heart rate does not vary appreciably over a period of time [74]. The parasympathetic dysfunction causes elevation of heart rate and in presence of sympathetic dysfunction heart rate may even decline [81]. Thus no rise in heart rate in presence of combined cardiovascular and somatic neuropathy complications could be attributed to involvement of sympathetic dysfunction in subjects with combined cardiovascular and somatic neuropathy [74].

A Brief over View of Treatment of Cardiac Autonomic Neuropathy (CAN)

Interventions to ameliorate reduced HRV are being evaluated in clinical trials based on theories of the pathogenesis of diabetic neuropathy. Intensive glycemic control is critical in preventing the onset and slowing the progression of diabetic autonomic neuropathy. The results of the Diabetes Control and Complications Trial shows that intensive glycemic control prevent the development of abnormal RR variation and slows the deterioration of autonomic dysfunction over time [82]. For persons with type 2 diabetes, intensive insulin therapy showed a small tendency for improved autonomic function, whereas deterioration was noted in the conventionally treated group [83]. Lipoic acid, an antioxidant that reduces free radical formation, appears to slow progression of CAN [84, 85]. For persons with type 2 diabetes, the improvement in CAN was seen after 4 months of treatment with an oral dosage of 800 mg/d [85]. Vitamin E has been shown to improve the ratio of cardiac sympathetic to parasympathetic tone for persons with type 2 diabetes [86]. Results of animal studies have suggested that impaired ganglion blood flow in diabetes could be responsible for neurodegenerative changes in autonomic postganglionic cell bodies [87]. However, with regard to changes in HRV, the use of ACE inhibitors in patients with CAN has resulted in conflicting outcomes. Of the ACE inhibitors studied, 12 months of use of quinapril showed some degree of success in treating CAN [88], whereas no improvement of cardiovascular autonomic function was shown after 12 months of treatment with trandolapril [89].

Aldosterone has been shown to affect the autonomic nervous system with sympathetic activation and parasympathetic inhibition [90]. Mechanisms thought to promote the beneficial effect of spironolactone include blocking the effect of aldosterone on the loss of potassium and magnesium and improved HRV [91-93].

It is well known that exercise plays an important role in the treatment of diabetes. The role of exercise in the improvement of cardiovascular autonomic function is not as clear. Endurance training was also shown to improve vagal activity in insulin-requiring diabetic individuals with early CAN [94]. But other studies showed no benefit or only minimal benefit for individuals with type 2 diabetes [95].

A number of different therapeutic agents are emerging for the treatment of diabetic neuropathy. Not all investigational drugs have, however, been studied with regard to the effect on autonomic nerve-fiber function. Nonetheless, given the multifactorial process involved in the pathogenesis of diabetic neuropathy, it is likely that combination therapies directed at various components of the pathogenic pathway may be required [96]

References

[1] Casellini, C. M., Vinik, A. I. Recent advances in the treatment of diabetic neuropathy. *Curr. Opin. Endocrinol. Diabetes* 2006; 13:147–153.

[2] Vinik, A. I., Ziegler, D. Diabetic Cardiovascular Autonomic Neuropathy. *Circulation*. 2007; 115:387-397.

[3] Schumer, M. P., Joyner, S. A., Pfeifer, M. A. Cardiovascular autonomic neuropathy testing in patients with diabetes. *Diabet. Spectr*. 1998; 11: 227–223.

[4] Nolan, R. P., Barry-Bianchi, S. M., Mechetiuc, A. E., Chen, M. H. Sex-based differences in the association between duration of type 2 diabetes and heart rate variability. *Diabetes and Vascular Disease Research* 2009; 6: 276–282.

[5] Vinik, A. I., Maser, R. E., Mitchell, B. D., Freeman, R. Diabetic autonomic neuropathy. *Diabetes Care* 2003; 26: 1553–579.

[6] Vinik, A. I., Erbas, T. Recognizing and treating diabetic autonomic neuropathy. *Cleveland clinic journal of medicine* 2011; 68:928:944.

[7] Stein, P. K., Bosner, M. S., Kleiger, R. E., Conger, B. M. Heart rate variability: A measure of cardiac autonomic tone. *Am. Heart J.* 1994; 127:1376-1381.

[8] Howorka, K., Pumprla, J., Schabmann, A. Optimal parameters of short-term heart rate spectrogram for routine evaluation of diabetic cardiovascular autonomic neuropathy. *J. Auton. Nerv. Syst.* 1998; 69: 104-172.

[9] Bigger, J. T. Jr, Fleiss, J. L., Rolnitzky, L. M., Steinman, R. C. Stability over time of heart period variability in patients with previous myocardial infarction and ventricular arrhythmias. The CAPS and ESVEM investigators. *Am. J. Cardiol.* 1992; 69:718-723.

[10] Akselrod, S., Gordon, D., Ubel, F. A., Shannon, D. C., Barger, A. C., Cohen, R. J. Power spectrum analysis of heart rate fluctuation: a quantitative probe of beat-to-beat cardiovascular control. *Science* 1981; 213:220-222.

[11] The Task Force of the European Society of Cardiology and the North American Society of Pacing Electrophysiology [ESC and NASPE]. Heart rate variability – standards of measurement, physiological interpretation and clinical use. *Circulation* 1996; 93:1043-1065.

[12] Pomeranz, B., Macaulay, R. J., Caudill, M. A., Kutz, I., Adam, D., Gordon, D., Kilborn, K. M., Barger, A. C., Shannon, D. C., Cohen, R. J., et al. Assessment of autonomic function in humans by heart rate spectral analysis. *Am. J. Physiol.* 1985; 248:H151-H153.

[13] Rimoldi, O., Pierini, S., Ferrari, A., Cerutti, S., Pagani, M., Malliani, A. Analysis of short-term oscillations of R-R and arterial pressure in conscious dogs. *Am. J. Physiol.* 1990; 258:H967-H976.

[14] Montano, N., Ruscone, T. G., Porta, A., Lombardi, F., Pagani, M., Malliani, A. Power spectrum analysis of heart rate variability to assess the changes in sympathovagal balance during graded orthostatic tilt. *Circulation* 1994; 90:1826-1831.

[15] Hayano, J., Sakakibara, Y., Yamada, A., Yamada, M., Mukai, S., Fujinami, T., Yokoyama, K., Watanabe, Y., Takata, K. Accuracy of assessment of cardiac vagal tone by heart rate variability in normal subjects. *Am. J. Cardiol.* 1991; 67:199-204.

[16] Malliani, A., Pagani, M., Lombardi, F., Cerutti, S. Cardiovascular neural regulation explored in the frequency domain. *Circulation* 1991; 84: 1482-1492.

[17] Hedman, A. E., Tahvanainen, K. U., Hartikainen, J. E., Hakumäki, M. O. Effect of sympathetic modulation and sympatho-vagal interaction on heart rate variability in anaesthetized dogs. *Acta Physiol. Scand.* 1995; 155:205-214.

[18] Pagani, M., Lombardi, F., Guzzetti, S., Guzzetti, S., Sandrone, G., Rimoldi, O., Malfatto, G., Cerutti, S., Malliani, A. Power spectral density of heart rate variability as an index of sympatho-vagal interaction in normal and hypertensive subjects. *J. Hypertens.* 1984;2 (suppl.):S383-S385.

[19] Kamath, M. V., Ghista, D. N., Fallen, E. L., Fitchett, D., Miller, D., McKelvie, R. Heart rate variability power spectrogram as a potential non-invasive signature of cardiac regulatory system response, mechanisms, and disorders. *Heart Vessels* 1987; 3:33-41.

[20] Sroka, K. Vagale Depression und ischämische Herzkrankheit. *Herz Kreisl.* 1998; 30:437-443.

[21] Stein, P. K., Kleiger, R. E. Insights from the study of heart rate variability. *Annu. Rev. Med.* 1999; 50:249-261.

[22] Bigger, J. T. Jr, Fleiss, J. L., Steinman, R. C., Rolnitzky, L. M., Kleiger, R. E., Rottman, J. N. Correlations among time and frequency domain measures of heart period variability two weeks after acute myocardial infarction. *Am. J. Cardiol.* 1992; 69:891-898.

[23] Dekker, J. M., Schouten, E. G., Klootwijk, P., Pool, J., Swenne, C. A., Kromhout, D. Heart rate variability from short electrocardiographic recordings predicts mortality from all causes in middle-aged and elderly men: The Zutphen Study, *Am. J. Epidemiol.* 1997; 145:899-908.

[24] Bruyne, M. C., Kors, J. A., Hoes, A. W., Klootwijk, P., Dekker, M. J., Hofman, A., Bemmel, J. H., Grobbee, D. E. Both decreased and increased heart rate variability on the standard 10-second electrocardiogram predict cardiac mortality in the elderly, *Am. J. Epidemiol.* 1999; 150: 1282-1288.

[25] Ewing, D. J., Clarke, B. F. Diagnosis and management of diabetic autonomic neuropathy. *Br. Med. J.* (Clin. Res. Ed.) 1982; 285:916-918.

[26] O'Brien, I. A., McFadden, J. P., Corrall, R. J. The influence of autonomic neuropathy on mortality in insulin-dependent diabetes. *Q J. Med.* 1991; 79:495-502.

[27] Navarro, X., Kennedy, W. R., Sutherland, D. E. Autonomic neuropathy and survival in diabetes mellitus: effects of pancreas transplantation. *Diabetologia* 1991; 34(suppl. 1):S108-112.

[28] Talman, W. T., Benaroch, E. E. Neural control of cardiac function. In: Dyck, P. J., Thomas, P. K., Lambert, E. H., Bunge, R., eds. *Peripheral Neuropathy*. Vol. 1. 3rd ed. Philadelphia: W.B Saunders, 1992: 177-185.

[29] Roy Freeman. Noninvasive evaluation of heart rate variability. In: Low, P. A., ed. Clinical autonomic disorders. *Evaluation and management.* Second edition. Philadelphia: Lippincot, 1997: 297-307.

[30] Weimer, L. H. Autonomic function. In: Evans, R. W., ed. *Diagnostic Testing in Neurology*. Philadelphia: W.B Saunders, 1999: 337-365.

[31] Genovely, H., Pfeifer, M. A. RR—variation: the autonomic test of choice in diabetes. *Diabetes Metab. Rev*. 1988; 4:255-271.

[32] Smith, S. A. Reduced sinus arrhythmia in diabetic autonomic neuropathy: diagnostic value of an age-related normal range. *BMJ*. 1982; 285:1599-1601.

[33] Engstrom, J. W., Martin, J. B. Disorders of the autonomic nervous system. In: Braunwald, E., Hauser, S. L., Fauci, A. S., et al., eds. *Harrison's Principles of Internal Medicine* (Vol. 2. 15th edition). New York: McGraw-Hill, 2001: 2416-2421.

[34] O'Brien, I. A., O'Hare, J. P., Lewin, I. G., Corrall, R. J. The prevalence of autonomic neuropathy in insulin-dependent diabetes mellitus: a controlled study based on heart rate variability. *Q J. Med*. 1986; 61:957-967.

[35] Low, P. A., Opfer-Gehrking, T. L., Proper, C. J., Zimmerman, I. The effect of aging on cardiac autonomic and postganglionic sudomotor function. *Muscle Nerve* 1990; 13:152-157.

[36] The DCCT Research Group. Factors in development of diabetic neuropathy. Baseline analysis of neuropathy in feasibility phase of Diabetes Control and Complications Trial (DCCT). *Diabetes* 1988; 37: 476–481.

[37] Jaffe, R. S., Aoki, T. T., Rohatsch, P. L., Disbrow, E. A., Fung, D. L. Predicting cardiac autonomic neuropathy in type 1 (insulin-dependent) diabetes mellitus. *Clin. Auton. Res*. 1995; 5:155-158.

[38] Maser, R. E., Pfeifer, M. A., Dorman, J. S., Kuller, L. H., Becker, D. J., Orchard, T. J. Diabetic autonomic neuropathy and cardiovascular risk.

Pittsburgh epidemiology of diabetes complications study III. *Arch. Intern. Med.* 1990; 150:1218-1222.

[39] Töyry, J. P., Niskanen, L. K., Mäntysaari, M. J., Länsimies, E. A., Uusitupa, M. I. Occurrence, predictors, and clinical significance of autonomic neuropathy in NIDDM. Ten-year follow-up from the diagnosis. *Diabetes* 1996; 45:308-315.

[40] Subbalakshmi, N. K., Adhikari, P. M. R., Rajeev, A., Asha, K., Jeganathan, P. S. Independent predictors of cardiac parasympathetic dysfunction in type 2 diabetes mellitus. *Singapore Med. J.* 2008; 49: 121-128.

[41] Dyrberg, T., Benn, J., Christiansen, S. J., Hilsted, J., Nerup, J. Prevalence of diabetic autonomic neuropathy measured by simple bedside tests. *Diabetologia* 1981; 20:190-194.

[42] Gundersen, H. J., Neubauer, B. A long-term diabetic autonomic nervous abnormality. Reduced variations in resting heart rate measured by a simple and sensitive method. *Diabetologia* 1977; 13:137-140.

[43] Straub, R. H., Zietz, B., Palitzsch, K. D., Schölmerich, J. Impact of disease duration on cardiovascular and pupillary autonomic nervous function in IDDM and NIDDM patients. *Diabetes Care* 1996; 19:960-967.

[44] Veglio, M., Carpano-Maglioli, P., Tonda, L., Quadri, R., Giannella, R., Rosa, C., Fonzo, D. Autonomic neuropathy in non-insulin dependent diabetic patients: correlation with age, sex, duration and metabolic control of diabetes. *Diabete Metab.* 1990; 16:200-206.

[45] Lehtinen, J. M., Uustiupa, M., Siitonen, O., Pyörälä, K. Prevalence of neuropathy in newly diagnosed NIDDM and nondiabetic control subjects. *Diabetes* 1989; 38:1307-1313.

[46] Pfeifer, M. A., Weinberg, C. R., Cook, D. L., Reenan, A., Halter, J. B., Ensinck, J. W., Porte, D. Jr. Autonomic neural dysfunction in recently diagnosed diabetic subjects. *Diabetes Care* 1984; 7:447-453.

[47] Spallone, V., Maiello, M. R., Cicconetti, E., Menzinger, G. Autonomic neuropathy and cardiovascular risk factors in insulin-dependent and non-insulin-dependent diabetes. *Diabetes Res. Clin. Pract.* 1997; 34:169-179.

[48] Valensi, P., Huard, J. P., Giroux, C., Attali, J. R. Factors involved in cardiac autonomic neuropathy in diabetic patients. *J. Diabetes Complications* 1997; 11:180-187.

[49] Mehta, S., Mathur, D., Chaturvedi, M., Verma, K. Incidence of cardiac autonomic neuropathy and its correlation with retinopathy,

microalbuminuria and glycated haemoglobin in non-insulin dependent diabetes mellitus. *J. Indian Med. Assoc.* 2002; 100:141-143.

[50] Lisa VanHoose, Youssef Sawers, Rajprasad Loganathan, James L Vacek, Lisa Stehno-Bittel, Lesya Novikova, Muhammed Al-Jarrah, Irina V Smirnova. Electrocardiographic changes with the onset of diabetes and the impact of aerobic exercise training in the Zucker Diabetic Fatty (ZDF) rat. *Cardiovascular Diabetology* 2010: 9:56.

[51] Gorodeski, E. Z., Ishwaran, H., Blackstone, E. H., Lauer, M. S. Quantitative electrocardiographic measures and long-term mortality in exercise test patients with clinically normal resting electrocardiograms. *Am. Heart J.* 2009; 158:61-70.

[52] Cardoso, C. R., Salles, G. F., Deccache, W. QTc interval prolongation is a predictor of future strokes in patients with type 2 diabetes mellitus. *Stroke* 2003; 34:2187-2194.

[53] Christensen, P. K., Gall, M. A., Major-Pedersen, A., Sato, A., Rossing, P., Breum, L., Pietersen, A., Kastrup, J., Parving, H. H. QTc interval length and QT dispersion as predictors of mortality in patients with non-insulin-dependent diabetes. *Scand. J. Clin. Lab. Invest.* 2000; 60:323-332.

[54] Veglio, M., Giunti, S., Stevens, L. K., Fuller, J. H., Perin, P. C. Prevalence of QT interval dispersion in type 1 diabetes and its relation with cardiac ischemia: the EURODIAB IDDM complications study group, *Diabetes Care* 2002; 25: 702-707.

[55] Veglio, M., Bruno, G., Borra, M., Macchia, G., Bargero, G., D'Errico, N., Pagano, G. F., Cavallo-Perin, P. Prevalence of increased QT interval duration and dispersion in type 2 diabetic patients and its relationship with coronary heart disease: a population-based cohort. *J. Intern. Med.* 2002; 251: 317-324.

[56] Veglio, M., Borra, M., Stevens, L. K., Fuller, J. H., Perin, P. C. The relation between QTc interval prolongation and diabetic complications: The EURODIAB IDDM complication study group, *Diabetologia* 1999; 42: 68-75.

[57] Rana, B. S., Lim, P. O., Naas, A. A., Ogston, S. A., Newton, R. W., Jung, R. T., Morris, A. D., Struthers, A. D. QT interval abnormalities are often present at diagnosis in diabetes and are better predictors of cardiac death than ankle brachial pressure index and autonomic function tests. *Heart* 2005; 91: 44-50.

[58] Subbalakshmi, N. K., Adhikari, P. M. R., Sathyanarayana Rao, K. N., Jeganathan, P. S. Influencing factors of qtc among the clinical

characteristics in type 2 diabetes mellitus. *Diabetes Research and Clinical Practice* 2010; 88:265-272.

[59] Moss, A. J. Measurement of the QT interval and the risk associated with QTc interval prolongation: A review, *Am. J. Cardiol.* 1993; 72: 23B:25 B.

[60] Abdülkadir, K., Zekeriya, A., Emin, M., Atilla, T. Orthostatic hypotension and heart rate variability as indicators of cardiac autonomic neuropathy in diabetes mellitus. *Eur. J. Gen. Med.* 2005; 2: 5-9.

[61] Katsuoka, H., Y. Mimori Y, Kurokawa K, Harada T, Kohriyama T, Ishizaki F, Harada A, Nakamura S. QTc interval and autonomic and somatic nerve function in diabetic neuropathy, *Clin. Auton. Res.* 1988; 8:39–143.

[62] Bellavere, F., Ferri, M., Guarini, L., Bax, G., Piccoli, A., Cardone, G., Piccoli, A., Cardone, C., Fedele, D. Prolonged QT period in diabetic autonomic neuropathy: a possible role in sudden cardiac death. *Br. Heart J.* 1988; 59: 379-383.

[63] Pappachan, J. M., Sebastian, J., Bino, B. C., Jayaprakash, K., Vijayakumar, K., Sujathan, P., Adinegara, L. A. Cardiac autonomic neuropathy in diabetes mellitus: prevalence, risk factors and utility of corrected QT interval in the ECG for its diagnosis. *Postgrad. Med. J.* 2008; 84:205-210.

[64] Low, P. A., Laboratory evaluation of autonomic function. In: P.A. Low, Editor, *Clinical Autonomic Disorders* (2nd Ed.). Philadelphia, Lippincott-Raven 1997; 179-208.

[65] Ko Gary, T. C., Juliana, C. N., Critchley Julian, A. J., Cockram Clive, S. Cardiovascular disease in Chinese type 2 diabetic women is associated with a prolonged QTc interval, *International journal of cardiology* 2000; 76: 75-80.

[66] Gastaldelli, A., Emdin, M., Conforti, F., Camastra, S., Ferrannini, E. Insulin prolongs the QTc interval in humans. *Am. J. Physiol. Regul. Integr. Comp. Physiol.* 2000; 279: R2022-2025.

[67] Bravenboer, B., Hendriksen, P. H., Oey, L. P., W. H. Gispen W. H., Van Huffelen, A. C., Erkelen, D. W. Is the corrected QT interval a reliable indicator of the severity of diabetic autonomic neuropathy? *Diabetes Care* 1993; 16:1249-1253.

[68] Barathwa, S. P. L., Agarwal, R., Khanna, D., Kumar, P. QTc prolongation in diabetes mellitus-an indicator of cardiac autonomic neuropathy. *JAPI* 1997; 45:15-17.

[69] Ahnve, S., Vallin, H. Influence of heart rate and inhibition of autonomic tone on the Q -T interval, *Circulation* 1981; 65: 435- 439.

[70] Karamitsos, D. T., Didangelos, T. P., Athyros, V. G., Kontopoulos, A. G. The natural history of recently diagnosed autonomic neuropathy over a period of 2 years. *Diab. Res. Clin. Pract*. 1998; 42:55–63.

[71] Mackay, J. D., Page, M. M., Cambridge, J., Watkins, P. J. Diabetic autonomic neuropathy. The diagnostic value of heart rate monitoring. *Diabetologia* 1980; 18:471–478.

[72] Progression of diabetic autonomic neuropathy over a decade in insulin-dependent diabetics. *QJM* 1990; 75:635–646.

[73] Dyck, P. J., Kratz, K. M., Karnes, J. L., Litchy, W. J., Klein, R., Pach, J. M., Wilson, M., Karnes, J. L., O'Brien, P. C., Melton, L. J. The prevalence by staged severity of various types of diabetic neuropathy, retinopathy, and nephropathy in a population-based cohort: the Rochester Diabetic Neuropathy Study. *Neurology* 1993; 43: 817–824.

[74] Subbalakshmi, N. K., Adhikari, P. M. R., Sathyanarayana Rao, K. N., Jeganathan, P. S. Deterioration of cardiac autonomic function over a period of one year in relation to cardiovascular and somatic neuropathy complications in type 2 diabetes mellitus. *Diabetes research and clinical practice* 2012; 97:313-321.

[75] Sundkvist, G., Almer, L-O., Lilja, B. Respiratory influence on heart rate in diabetes mellitus. *Br. Med. J*. 1979; 1:924–925.

[76] Kleiger, R. E., Stein, P. K., Bosner, M. S., Rotman, J. N. Time domain measurements of heart rate variability. *Cardiol. Clin*. 1992; 10:487–498.

[77] Sztajzel, J. Heart rate variability: a noninvasive electrocardiographic method to measure the autonomic nervous system. *Swiss Med. Wkly* 2004; 134:514–522.

[78] Moser, M., Lehofer, M., Sedminek, A., Lux, M., Zapotoczky, H. G., Kenner, T., Noordergraaf, A. Heart rate variability as a prognostic tool in cardiology. A contribution to the problem from a theoretical point of view. *Circulation* 1994; 90: 1078–1082.

[79] Schwartz, P., Larovere, M. T., Vanoli, E. Autonomic nervous system and sudden death. Experimental basis and clinical observations for post-myocardial risk stratification. *Circulation* 1992; 85:177–191.

[80] Takahashi, N., Nakagawa, M., Saikawa, T., Ooie, T., Yufu, K., Shigematsu, S., Hara, M., Sakino, H., Katsuragi, I., Okeda, T., Yoshimatsu, H., Sakata, T. Effect of essential hypertension on cardiac autonomic function in type 2 diabetic patients. *J. Am. Coll. Cardiol*. 2001; 38:232–237.

[81] Ewing, D. J., Cambell, I. W., Clarke, B. F. Heart rate changes in diabetes mellitus. *Lancet* 1981; 183–185.

[82] The Diabetes Control and Complications Trial Research Group. The effect of intensive diabetes therapy on measures of autonomic nervous system function in the Diabetes Control and Complications Trial (DCCT). *Diabetologia* 1998; 41:416–423.

[83] Ohkubo, Y., Kishikawa, H., Araki, E., Miyata, T., Isami, S., Motoyoshi, S., Kojima, Y., Furuyoshi, N., Shichiri, M. Intensive insulin therapy prevents the progression of diabetic microvascular complications in Japanese patients with non-insulin-dependent diabetes mellitus: a randomized prospective 6-year study. *Diabetes Res. Clin. Pract.* 1995; 28:103–117.

[84] Tankova, T., Koev, D., Dakovska, L. Alpha-lipoic acid in the treatment of autonomic diabetic neuropathy (controlled, randomized, open-label study). *Rom. J. Intern. Med.* 2004; 42:457–464.

[85] Ziegler, D., Schatz, H., Conrad, F., Gries, F. A., Ulrich, H., Reichel, G. Effects of treatment with the antioxidant alpha-lipoic acid on cardiac autonomic neuropathy in NIDDM patients. A 4-month randomized controlled multicenter trial (DEKAN Study). *Diabetes Care* 1997; 20: 369–373.

[86] Manzella, D., Barbieri, M., Ragno, E., Paolisso, G Chronic administration of pharmacologic doses of vitamin E improves the cardiac autonomic nervous system in patients with type 2 diabetes. *Am. J. Clin. Nutr.* 2001; 73:1052–1057.

[87] Cameron, N. E., Cotter, M. A. Diabetes causes an early reduction in autonomic ganglion blood flow in rats. *J. Diabetes Complications* 2001; 15:198–202.

[88] Athyros, V. G., Didangelos, T. P., Karamitsos, D. T., Papageorgiou, A. A., Boudoulas, H., Kontopoulos, A. G. Long-term effect of converting enzyme inhibition on circadian sympathetic and parasympathetic modulation in patients with diabetic autonomic neuropathy. *Acta Cardiol.* 1998; 53:201–209.

[89] Malik, R. A., Williamson, S., Abbott, C., Carrington, A. L., Iqbal, J., Schady, W., Boulton, A. J. Effect of angiotensin-converting-enzyme (ACE) inhibitor trandolapril on human diabetic neuropathy: randomised double-blind controlled trial. *Lancet* 1998; 352:1978–1981.

[90] Pitt, B., Zannad, F., Remme, W. J., Cody, R., Castaigne, A., Perez, A., Palensky, J., Wittes, J., for the Randomized Aldactone Evaluation Study

Investigators The effect of spironolactone on morbidity and mortality in patients with severe heart failure. *N Engl. J. Med.* 1999; 341:709–717.

[91] MacFadyen, R. J., Barr, C. S., Struthers, A. D. Aldosterone blockade reduces vascular collagen turnover, improves heart rate variability and reduces early morning rise in heart rate in heart failure patients. *Cardiovasc. Res.* 1997; 35:30–34.

[92] Korkmaz, M. E., Muderrisoglu, H., Ulucam, M., Ozin, B. Effects of spironolactone on heart rate variability and left ventricular systolic function in severe ischemic heart failure. *Am. J. Cardiol.* 2000; 86:649–653.

[93] Yee, K. M., Pringle, S. D., Struthers, A. D. Circadian variation in the effects of aldosterone blockade on heart rate variability and QT dispersion in congestive heart failure. *J. Am. Coll. Cardiol.* 2001; 37: 1800–1807.

[94] Howorka, K., Pumprla, J., Haber, P., Koller-Strametz, J., Mondrzyk, J., Schabmann, A. Effects of physical training on heart rate variability in diabetic patients with various degrees of cardiovascular autonomic neuropathy. *Cardiovasc. Res.* 1997; 34:206–214.

[95] Loimaala, A., Huikuri, H. V., Koobi, T., Rinne, M., Nenonen, A., Vuori, I. Exercise training improves baroreflex sensitivity in type 2 diabetes. *Diabetes* 2003; 52:1837–1842.

[96] Maser, R. E. and James Lenhard, M. REVIEW: Cardiovascular Autonomic Neuropathy. Due to Diabetes Mellitus: Clinical Manifestations, Consequences, and Treatment. *J. Clin. Endocrinol. Metab.* 2005; 90:5896–5903.

In: Advances in Neuropathy Research
Editors: M. O. Brown and B. A. Budd
ISBN: 978-1-62618-400-8

Chapter 3

Objective Investigation of Multiple Sclerosis Related Optic Neuritis. What's New?

Irini Chatziralli*[1,2*]*, Marilita Moschos*[1]*,
***Dimitrios Brouzas*[1]*, Vasileios Peponis*[2]
***and Ioannis Ladas*[1]

[1]Department of Ophthalmology, University of Athens, Athens, Greece
[2]Eye Clinic, Specialized Eye Hospital "Ophthalmiatrion Athinon", Athens, Greece

Abstract

Optic neuritis (ON) is an inflammatory, infective or demyelinating process, affecting the optic nerve. Retrobulbar neuritis, in which the optic disc appearance is normal, at least initially, is the most frequent type in adults and is often associated with multiple sclerosis. Multiple sclerosis is a complex neurodegenerative disorder which can cause visual dysfunction in 80% of patients during the course of the disease. Axonal loss is considered to be responsible for visual disability in ON associated with MS rather than demyelination with preservation of axons. The value of magnetic resonance imaging (MRI) in the diagnosis of multiple

* Corresponding author: Irini Chatziralli MD; 28, Papanastasiou street, Agios Dimitrios, 17342, Athens, Greece; Tel: 00306973046326; Fax: 00302109312008; E-mail: eirchat@yahoo.gr.

sclerosis is well established, as MRI provides information about disease burden, emphasizing on inflammation and demyelination. However, there is a poor correlation between quantification of axonal and neuronal loss by MRI and functional disabilities evaluated by clinical measures.

Optical coherence tomography (OCT) is a non-invasive high-resolution method, routinely used in ophthalmology to visualize and quantify the layers of the retina. It provides information on optic nerve head topography, retinal nerve fiber layer (RNFL) thickness and macular volume, which correlate with axonal loss. Several studies found a significant reduction in RNFL thickness in eyes affected by multiple sclerosis with or without optic neuritis, compared to control eyes. RNFL thickness was correlated with visual acuity, visual field, contrast sensitivity and colour vision.

In addition to OCT, which provides quantitative measurements of the RNFL thickness, visual evoked potentials (VEP) offer a functional assessment of the integrity of the optic nerve in ON. OCT findings were correlated with amplitude of visual evoked potentials (VEP), but not with latency. Nowadays the new multifocal VEP (mVEP) allows the detection of multiple individual VEP signals across the visual field up to eccentricity of 20° to 30°. As the changes of optic nerve are more diffuse than originally thought based on standard perimetry, mVEP was found to be helpful in predicting progression to future multiple sclerosis.

Introduction

Optic neuritis (ON) is an inflammatory optic nerve injury, which causes acute to subacute onset vision loss in children and young adults [1,2]. Over 90% of patients report pain at the onset of vision loss, which worsens with eye movements [1]. Vision loss usually progresses over a period of hours to days and may range from mild to "no light perception" [2]. Decreased color vision is quite common, while patients often manifest a relative afferent pupil defect, unless there is coexisting optic nerve damage in the contralateral eye [1,2].

ON can be retrobulbar, in which the optic disc appearance is normal, at least initially, or anterior ON, presenting with optic disc swelling [2]. The majority of ON patients recover vision over a period of weeks, during which time optic disc pallor may evolve on a "footprint" of the previous inflammation. Nevertheless, visual problems, such as loss of contrast sensitivity, loss of depth perception and fatigue-induced vision loss, are common. ON and especially retrobulbar ON is thought to be associated with multiple sclerosis (MS) [1-7]. Multiple sclerosis is a complex neurodegenerative disorder, characterized by inflammation, demyelination and axonal loss, the latter being considered as the pathological correlation of clinical disability [4,6,7]. It is considered that an autoimmune attack to the myelin sheath occurs and leads to tissue damage, often involving the prechiasmatic visual pathways [5]. MS can cause visual dysfunction in 80% of patients during the course of the disease and approximately 15-20% of MS patients present with ON, while ON occurs at some point in 50% of patients with established MS [2,8,9]. Interestingly, axonal loss is considered to be responsible for visual disability in ON associated with MS rather than demyelination with preservation of axons [5,8,9].

The value of magnetic resonance imaging (MRI) in the diagnosis of MS is well established, as MRI gives information about inflammation and demyelination and therefore about disease severity [10]. However, conventional MRI provides insufficient sensitivity and specificity to reveal the degree of damage to the central nervous system [3]. Additionally, there is a poor correlation between quantification of axonal and neuronal loss by MRI and functional disabilities evaluated by clinical measures [10,11].

In contrast to MRI, peripapillary retinal nerve fibre layer (RNFL) thickness is suggested as a structural marker of axonal loss, because the unmyelinated axons can be imaged directly and neurodegeneration can be quantified in vivo [11-14]. Besides optical coherence tomography (OCT), which provides quantitative measurements of the RNFL thickness, visual evoked potentials (VEP) offer an objective functional assessment of the integrity of the optic nerve in ON [15]. In light of the above, herein our purpose is to summarize the results of published studies, evaluating the new methods of objective investigation of ON associated with MS (MSON).

Optical Coherence Tomography

The RNFL is devoid of myelin and contains the axons of retinal ganglion cells (RGC). The axons converge at the optic nerve head, where they form the optic nerve. The acute lesion affecting the optic nerve in ON could therefore cause atrophy of the RNFL by acute transection of axons, and consequently by retrograde degeneration. This assertion is supported by data from primate studies where optic nerve transection leads to retrograde loss of RGC bodies [16].

Optical coherence tomography is a noninvasive tool, providing an "optical biopsy" of the retina and quantifying RNFL thickness in vivo. OCT has been considered as a highly sensitive method for imaging of neurodegeneration in

MS, as the retina is the only place where axons can be imaged directly [17]. Thus, quantification of the retinal nerve fiber layer (RNFL) enables the monitoring of degeneration [18]. Axonal loss is not reversible in contrast to demyelination and it is thought to be a significant cause of sustained disability. As a result, a validated tool for monitoring axonal loss is needed [18]. Post-mortem studies have revealed that atrophy of RNFL and RGC layer is present in the majority of MS patients even without a clinical history of ON [19].

In MSON there is a loss of RNFL thickness ranged from 5 to 40 μm with an average at 10-20 μm [1]. This loss was significant in all studies using time domain OCT (TD-OCT). Parisi et al first studied a small cohort of patients with MS and relatively good visual function. They found that RNFL thickness was decreased, even though the patients exhibited visual recovery [20]. On the other hand, Trip et al reported retrospective data in 25 patients with single attack of ON and poor visual recovery, excluding patients with acute ON. In this study, the mean RNFL thickness and macular volume was reduced in the affected eyes of patients compared to the unaffected eyes and to control eyes [16]. A nonsignificant reduction in unaffected eyes compared to control eyes was found in this study, which again was of lower magnitude to the findings of Parisi and colleagues [16,20]. Fellow eye abnormalities do occur in unilateral ON, and a possible explanation for the difference between the two study findings (as well as technical differences relating to the OCT model used) is that the unaffected eyes in this study had a lesser degree of subclinical involvement than those in the study by Parisi and colleagues, all of whom had clinically definite MS. Of note, reduced RNFL thickness predicted significantly worse function for the visual acuity, the visual field and the color vision [16,20].

One year later, Fisher et al compared RNFL thickness in 90 MS patients versus 36 controls and found, in line with Albrecht et al and Cheng et al, that it was reduced with the lowest volume noted in the ON eyes of MS patients [3,11,21].

Noticeably, according to Albrecht et al, RNFL thickness may correlate with progressive course of disease and with clinical burden of the disease [3]. Moreover, axonal loss measured by OCT correlated with neuronal degeneration in the brain. Thus, OCT can be a useful biomarker to monitor axonal loss and neuronal degeneration in MS, as Gundogan et al also supported [14]. Several other studies shared the same results using TD-OCT [22-43], as well as the new spectral domain OCT (SD-OCT), which provide a more accurate description of retinal layers [44-49]. As a rule of thumb, RNFL loss becomes readily detectable with OCT about 3 months after acute ON.

Early reduction in RNFL thickness caused by axonal atrophy is clinically difficult to distinguish from a reduction due to resolution of axonal swelling, which is common in acute ON [34,50].

Costello et al presented longitudinal data for RNFL thickness during the first year after ON, where continuous axonal loss in the affected eye occurs for at least 12 months, but the most thinning is present 6 months after ON, in line with Chatziralli et al, who also found a progressive decrease in RNFL thickness, which remained even though VA improved [34,39]. Another interesting finding of this study is that in MS eyes without ON, RNFL thickness was also abnormal [39].

This is in accordance with Talman et al, who demonstrated that there is progressive RNFL thinning as a function of time in MS generally [51]. Therefore, rapid axonal loss may occur in the affected eyes and the clinically unaffected eyes may have subclinical loss even in the absence of clinical symptoms, suggesting an inverse correlation between disease duration and average RNFL thickness [39,51]. Additionally, Burkholder et al showed that lower macular volumes were associated with RNFL thinning, so that a 10 μm difference in RNFL thickness corresponded to 0.20 mm2 reduction in total macular volume. Noticeably, correlation between RNFL thickness and inner macular volume were significant, especially in ON eyes compared to non-ON eyes [22]. Taken as a whole, OCT enables RNFL measurement and therefore direct visualization and quantification of the atrophy of unmyelinated axons in the eyes with a history of ON [1,3,11-18, 20-51]. It can also detect subclinical atrophy, as RNFL thinning occurs in eyes not previously affected by ON [23,24]. Moreover, OCT can be used to study the timing of RNFL changes and the correlation with visual function at various recovery stages of ON over a 12-month period [34,39], as well as help distinguish ON from neuromyelitis optica [32]. It is also worthy to note that OCT can be applied to quantify macular volume and RGC loss. Analogous to the grey matter atrophy in the MS brain, neuronal loss in the macula occurs with RNFL axonal degeneration [22]. Nevertheless, OCT cannot predict the future risk of MS [30].

Electrophysiological Tests (Electroretinogram, Visual Evoked Potentials, Multifocal Visual Evoked Potentials)

Paraclinical tests, including electrophysiological ones, are performed to increase the certainty of a clinically suspected diagnosis, and this is of particular importance in a disease of "no better explanation," such as multiple sclerosis (MS), for which no specific test exists [52]. To assess retinal function in MS patients, the electroretinogram (ERG) can be used, as MS is thought to be associated with retinal abnormalities [53]. The full-field flash ERG is a noninvasive test, where massed electrical potentials from the entire retina are recorded and signals from different layers of the retina can be evaluated [54]. The flash ERG is used to detect diseases affecting the photoreceptors and bipolar cells, whose signals are recorded by a- and b-wave respectively [53]. Accordingly, the activity of retinal ganglion cells is reflected by the photopic negative response (PhNR) [53,55]. Several studies have found reduced amplitude of the pattern ERG in MS patients, particularly the N95 component of the transient pattern ERG [20,56]. Wang et al reported significant reduction in PhNR amplitude even in unaffected fellow eyes of MS ON eyes and in MS patients with no history of ON in either eye [53]. In this study, a- and b- wave amplitudes were not significantly different in MS eyes than in controls, in line with previous studies, which have found that there was no difference in a-wave amplitude and latency between MS patients and normal subjects, as well as between affected and non-affected eyes [53,57,58]. As a result, subclinical pathologic changes associated with the disease may imply a potential cause of chronic damage to retinal neurons and their axons in MS patients [53].

The clinical utility of visual evoked potentials (VEPs), first employed by Dawson in 1947, is based on their ability to demonstrate objectively abnormal sensory system conduction, to reveal subclinical involvement, to help define the anatomic distribution, to give some insights into pathophysiology of a disease, and to monitor the changes in neurological status [59]. Since the first reports by Halliday et al in the early 1970s [60-62], the changes in pattern reversal VEPs in ON and MS have been documented [39,63-76]. Although most of the various assessment techniques are predominantly used in research, the conventional visual evoked potentials (cVEPs) are the most commonly employed electrophysiological tests for early diagnosis of demyelinating diseases in clinical practice [77]. Interestingly, Balnyte et al revealed a high diagnostic value of cVEPs in terms of both sensitivity and specificity (90.5% and 82.5%) for the MS patients when employed in everyday clinical practice [62].

In MS, it is very important to diagnose the disease early for choosing the optimal time of treatment initiation, as "time is also brain" [78]. Demyelination has a predilection to occur in certain sites within the central nervous system (CNS) [62]. The afferent visual pathway, which extends from the retina to the primary visual cortex and consists of four neuronal components, is often affected even without a history of ON and/or visual symptoms [62,79]. Since a few diseases other than MS affect the CNS, the demonstration of an optic nerve lesion helps define the disease. The electrophysiological assessment of the clinically unaffected eye of patients with unilateral ON may reflect the status of normal-appearing white matter in the CNS [79]. Thus, abnormal VEPs in unaffected eyes provide evidence that might help identify dissemination in space and hence help diagnose MS [79,80]. Klistorner et al compared results from OCT and mfVEP in patients with unilateral ON and found that the mfVEP detected more abnormality than the OCT RNFL thickness in both affected eyes and fellow eyes [80]. Nevertheless, Samsen et al supported that VEPs are used to demonstrate the demyelination procedure in patients with ON, but cannot determine the susceptibility of the patients with acute ON to become MS [76].

Multiple sclerosis has long been regarded as only a demyelinating disease, but evidence now suggests widespread damage to axons [62]. Demyelination, as well as axonal loss, is reflected in the changes in the initial component of the VEP response affecting its latency, waveform configuration, or amplitude [62,79,81]. The amplitude and latency of the first positive peak, P100, were determined to assess electrical conduction in patients compared to normal subjects [67]. Latencies of main peaks were prolonged and response amplitudes were reduced in MS patients. Specifically, Basar-Eroglu et al found that alpha (7-12 Hz) components of evoked potentials were significantly reduced whereas theta (4-7 Hz) response was not altered [67]. Neurophysiological pathological evidence may improve, worsen, remain the same, or in small percentage of cases, return temporarily to normal in time [39,82]. VEP latencies were found to decrease significantly during the first two years with the most marked changes occurring between 3 to 6 months suggesting that recovery processes involving remyelination and possibly ion channel reorganization proceed for at least two years in the background of concurrent effects of demyelination and/or axonal degeneration also occurring in the fellow eye optic nerve initially masked by the recovery process, but gradually becoming more evident [83]. Thus, VEP measures can provide not only diagnostic, but also prognostic information during the evaluation of patients with MS [39,62].

The multifocal visual evoked potential (mfVEP) is a relatively new objective technique with increasing research interest, which has to be established as a reliable diagnostic tool, assessing early visual pathway integrity [84]. It has been developed to examine conduction in the parts of the visual field that is not possible with the full field cVEPs [85-88]. This noninvasive electrodiagnostic technique records many (typically 60) local visual evoked responses simultaneously from over 40 degree field of vision. Additionally, the mfVEP provides information about nerve conduction velocity (latency), which is useful to assess the extent of demyelination [89]. The mfVEP has been found to have good repeatability, detecting local defects not being possible to find be means of the traditional VEP [90]. According to Grover et al, it was found that mfVEPs were more sensitive than cVEPs in patients with MS; however, the difference was less significant than expected, and thus, it seemed advisable firstly to test patients suspected of MS with cVEPs [77].

In addition, Hood et al noted two additional reasons why the mfVEP might be superior to the cVEP [91,92]. The cVEP records responses from a wide region of a field and local delays may be lost in the process [93]. Moreover, the cVEP is less likely to detect upper field defects, as the lower field can contribute more to the cVEP than the upper field [89,91,94]. As a result, although cVEP may detect wide-spread areas of demyelinization, it fails to detect localized defects, in contrast to mfVEP [90-94].

Conclusion

Multiple sclerosis is a complex neurodegenerative disorder which can cause visual dysfunction in 80% of patients during the course of the disease. Axonal loss is considered to be responsible for visual disability in ON associated with MS rather than demyelination with preservation of axons. The value of magnetic resonance imaging (MRI) in the diagnosis of multiple sclerosis is well established, as MRI provides information about disease burden, emphasizing on inflammation and demyelination. However, there is a poor correlation between quantification of axonal and neuronal loss by MRI and functional disabilities evaluated by clinical measures.

Optical coherence tomography enables RNFL measurement and therefore direct visualization and quantification of the atrophy, even subclinical, of unmyelinated axons in the MS eyes. Moreover, OCT can be used to study the timing of RNFL changes and the correlation with visual function at various recovery stages of ON over time, although it cannot predict the future risk of MS. On the other hand, VEP offer a functional assessment of the integrity of the optic nerve in ON. OCT findings were correlated with amplitude of visual evoked potentials (VEP), but not with latency. Nowadays the mfVEP allows the detection of multiple individual VEP signals across the visual field up to eccentricity of 20° to 30°. As the changes of optic nerve are more diffuse than originally thought based on standard perimetry, mVEP was found to be helpful in predicting progression to future MS.

References

[1] Costello, F; Coupland, S; Hodge, W; Lorello, GR; Koroluk, J; Pan, YI; Freedman, MS; Zackon, DH; Kardon, RH. Quantifying axonal loss after optic neuritis with optical coherence tomography. *Ann. Neurol.* 2006; 59, 963–969.

[2] Kanski, JJ. Neuro-ophthalmology. In: Elsevier Butterworth Heinemann, ed. Clinical Ophthalmology-A systematic approach. 6th ed., 2007, p 788-789.

[3] Albrecht, P; Fröhlich, R; Hartung, HP; Kieseier, BC, Methner, A. Optical coherence tomography measures axonal loss in multiple sclerosis independently of optic neuritis. *J. Neurol.* 2007; 254, 1595-1596.

[4] Frohman, EM; Racke, MK; Raine, CS. Multiple sclerosis--the plaque and its pathogenesis. *N. Engl. J. Med. 2006*; 354, 942-955.

[5] Iester, M; Cioli, F, Uccelli, A; Papadia, M; Bandini, F; Mancardi, GL; Calabria, GA. Retinal nerve fibre layer measurements and optic nerve head analysis in multiple sclerosis patients. *Eye* 2009; 23, 407-412.

[6] Compston, A; Coles, A. Multiple sclerosis. *Lancet* 2002; 359, 1221–1231.

[7] Murray, TJ. Diagnosis and treatment of multiple sclerosis. *BMJ* 2006; 332, 525-527.

[8] Sørensen, TL; Fredriksen, JL; Brønnum-Hansen, H; Petersen, HC. Optic neuritis as onset manifestation of multiple sclerosis: a nationwide, long-term survey. *Neurology* 1999; 53, 473-478.

[9] Plant, GT. Optic neuritis and multiple sclerosis. *Curr. Opin. Neurol.* 2008; 21, 16-21.

[10] Trip, SA; Miller, DH. Imaging in multiple sclerosis. *J. Neurol. Neurosurg. Psychiatry* 2005; 76, 11-18.

[11] Fisher, JB; Jacobs, DA; Markowitz, CE; Galetta, SL; Volpe, NJ; Nano-Schiavi, ML; Baier, ML; Frohman, EM; Winslow, H; Frohman, TC; Calabresi, PA; Maguire, MG; Cutter, GR; Balcer, LJ. Relation of visual function to retinal nerve fiber layer thickness in multiple sclerosis. *Ophthalmology* 2006; 113, 324-332.

[12] Kallenbach, K; Frederiksen, J. Optical coherence tomography in optic neuritis and multiple scleroses: a review. *Eur. J. Neurol.* 2007; 14, 841-849.

[13] Trip, SA; Schlottmann, PG; Jones, SJ; Li, WY; Garway-Heath, DF; Thompson, AJ; Plant, GT; Miller, DH. Optic nerve atrophy and retinal nerve fibre layer thinning following optic neuritis: evidence that axonal loss is a substrate of MRI-detected atrophy. *Neuroimage* 2006; 31, 286-293.

[14] Gundogan, FC; Demirkaya, S; Sobaci, G. Is optical coherence tomography really a new biomarker candidate in multiple sclerosis?--A structural and functional evaluation. *Invest. Ophthalmol Vis. Sci.* 2007; 48, 5773-5581.

[15] Klistorner, A; Arvind, H; Nguyen, T; Garrick, R; Paine, M; Graham, S; O'Day, J,; Yiannikas, C. Multifocal VEP and OCT in optic neuritis: a topographical study of the structure-function relationship. *Doc. Ophthalmol.* 2009; 118, 129-137.

[16] Trip, SA; Schlottmann, PG; Jones, SJ; Altmann, DR; Garway-Heath, DF; Thompson, AJ; Plant, GT; Miller, DH. Retinal nerve fiber layer axonal loss and visual dysfunction in optic neuritis. *Ann. Neurol.* 2005; 58, 383-391.

[17] Frohman, E; Costello, F; Zivadinov, R; Stuve, O; Conger, A; Winslow, H; Trip, A; Frohman, T; Balcer, L. Optical coherence tomography in multiple sclerosis. *Lancet Neurol.* 2006; 5, 853-863.

[18] Petzold, A; de Boer, JF; Schippling, S; Vermersch, P; Kardon, R; Green, A; Calabresi, PA; Polman, C. Optical coherence tomography in multiple sclerosis: a systematic review and meta-analysis. *Lancet Neurol.* 2010; 9, 921-932.

[19] Green, AJ; McQuaid, S; Hauser, SL; Allen, IV; Lyness, R. Ocular pathology in multiple sclerosis: retinal atrophy and inflammation irrespective of disease duration. *Brain* 2010; 133, 1591-1601.

[20] Parisi, V; Manni, G; Spadaro, M; Colacino, G; Restuccia, R; Marchi, S; Bucci, MG; Pierelli, F. Correlation between morphological and functional retinal impairment in multiple sclerosis patients. *Invest. Ophthalmol Vis. Sci.* 1999; 40, 2520-2527.

[21] Cheng, H; Laron, M; Schiffman, JS; Tang, RA; Frishman, LJ. The relationship between visual field and retinal nerve fiber layer measurements in patients with multiple sclerosis. *Invest. Ophthalmol Vis. Sci* 2007; 48, 5798-5805.

[22] Burkholder, BM; Osborne, B; Loguidice, MJ; Bisker, E; Frohman, TC; Conger, A; Ratchford, JN; Warner, C; Markowitz, CE; Jacobs, DA; Galetta, SL; Cutter, GR; Maguire, MG; Calabresi, PA; Balcer, LJ; Frohman, EM. Macular volume determined by optical coherence tomography as a measure of neuronal loss in multiple sclerosis. *Arch. Neurol.* 2009; 66, 1366-1372.

[23] Pulicken, M; Gordon-Lipkin, E; Balcer, LJ; Frohman, E; Cutter, G; Calabresi, PA. Optical coherence tomography and disease subtype in multiple sclerosis. *Neurology* 2007; 69, 2085-2092.

[24] Henderson, AP; Trip, SA; Schlottmann, PG; Altmann, DR; Garway-Heath, DF; Plant, GT; Miller, DH. An investigation of the retinal nerve fibre layer in progressive multiple sclerosis using optical coherence tomography. *Brain* 2008; 131, 277-287.

[25] Pueyo, V; Martin, J; Fernandez, J; Almarcegui, C; Ara, J; Egea, C; Pablo, L; Honrubia, F. Axonal loss in the retinal nerve fiber layer in patients with multiple sclerosis. *Mult. Scler.* 2008; 14, 609-614.

[26] Zaveri, MS; Conger, A; Salter, A; Frohman, TC; Galetta, SL; Markowitz, CE; Jacobs, DA; Cutter, GR; Ying, GS; Maguire, MG; Calabresi, PA; Balcer, LJ; Frohman, EM. Retinal imaging by laser polarimetry and optical coherence tomography evidence of axonal degeneration in multiple sclerosis. *Arch. Neurol.* 2008; 65, 924-928.

[27] Bock, M; Brandt, AU; Dörr, J; Kraft, H; Weinges-Evers, N; Gaede, G; Pfueller, CF; Herges, K; Radbruch, H; Ohlraun, S; Bellmann-Strobl, J; Kuchenbecker, J; Zipp, F; Paul, F. Patterns of retinal nerve fiber layer loss in multiple sclerosis patients with or without optic neuritis and glaucoma patients. *Clin. Neurol. Neurosurg.* 2010; 112, 647-652.

[28] Garcia-Martin, E; Pueyo, V; Martin, J; Almarcegui, C; Ara, JR; Dolz, I; Honrubia, FM; Fernandez, FJ. Progressive changes in the retinal nerve fiber layer in patients with multiple sclerosis. *Eur. J. Ophthalmol.* 2010; 20, 167-173.

[29] Merle, H; Olindo, S; Donnio, A; Beral, L; Richer, R; Smadja, D; Cabre, P. Retinal nerve fiber layer thickness and spatial and temporal contrast sensitivity in multiple sclerosis. *Eur. J. Ophthalmol.* 2010; 20, 158-166.

[30] Costello, F; Hodge, W; Pan, YI; Eggenberger, E; Coupland, S; Kardon, RH. Tracking retinal nerve fiber layer loss after optic neuritis: a prospective study using optical coherence tomography. *Mult. Scler.* 2008; 14, 893-905.

[31] Siger,M; Dziegielewski, K; Jasek, L; Bieniek, M; Nicpan, A; Nawrocki, J; Selmaj, K. Optical coherence tomography in multiple sclerosis:

thickness of the retinal nerve fiber layer as a potential measure of axonal loss and brain atrophy. *J. Neurol.* 2008; 255, 1555-1560.

[32] Noval, S; Contreras, I; Muñoz, S; Oreja-Guevara, C; Manzano, B; Rebolleda, G. Optical coherence tomography in multiple sclerosis and neuromyelitis optica: an update. *Mult. Scler. Int.* 2011; 2011, 472790.

[33] Tátrai, E; Simó, M; Iljicsov, A; Németh, J; Debuc, DC; Somfai GM. In vivo evaluation of retinal neurodegeneration in patients with multiple sclerosis. *PLoS One* 2012; 7,e30922.

[34] Costello, F. Evaluating the use of optical coherence tomography in optic neuritis. *Mult. Scler. Int.* 2011; 2011, 148394.

[35] Costello, FE; Klistorner, A; Kardon, R. Optical coherence tomography in the diagnosis and management of optic neuritis and multiple sclerosis. *Ophthalmic Surg. Lasers Imaging* 2011; 42, S28-40.

[36] Kitsos, G; Detorakis, ET; Papakonstantinou, S; Kyritsis, AP; Pelidou, SH. Perimetric and peri-papillary nerve fibre layer thickness findings in multiple sclerosis. *Eur. J. Neurol.* 2011; 18, 719-725.

[37] Pfueller, CF; Brandt, AU; Schubert, F; Bock, M; Walaszek, B; Waiczies, H; Schwenteck, T; Dörr, J; Bellmann-Strobl, J; Mohr, C; Weinges-Evers, N; Ittermann, B; Wuerfel, JT; Paul, F. Metabolic changes in the visual cortex are linked to retinal nerve fiber layer thinning in multiple sclerosis. *PLoS One* 2011; 6, e18019.

[38] Siepman, TA; Bettink-Remeijer, MW; Hintzen, RQ. Retinal nerve fiber layer thickness in subgroups of multiple sclerosis, measured by optical coherence tomography and scanning laser polarimetry. *J. Neurol.* 2010; 257, 1654-1660.

[39] Chatziralli, IP; Moschos, MM; Brouzas, D; Kopsidas, K; Ladas, ID. Evaluation of retinal nerve fibre layer thickness and visual evoked potentials in optic neuritis associated with multiple sclerosis. *Clin. Exp. Optom.* 2012; 95, 223-228.

[40] Quelly, A; Cheng, H; Laron, M; Schiffman, JS; Tang, RA. Comparison of optical coherence tomography and scanning laser polarimetry measurements in patients with multiple sclerosis. *Optom. Vis. Sci.* 2010; 87, 576-584.

[41] Naismith, RT; Tutlam, NT; Xu, J; Shepherd, JB; Klawiter, EC; Song, SK; Cross, AH. Optical coherence tomography is less sensitive than visual evoked potentials in optic neuritis. *Neurology* 2009; 73, 46-52.

[42] Grazioli, E; Zivadinov, R; Weinstock-Guttman, B; Lincoff, N; Baier, M; Wong, JR; Hussein, S; Cox, JL; Hojnacki, D; Ramanathan, M. Retinal

nerve fiber layer thickness is associated with brain MRI outcomes in multiple sclerosis. *J. Neurol. Sci.* 2008; 268, 12-17.

[43] Outteryck, O; Zephir, H; Defoort, S; Bouyon, M; Debruyne, P; Bouacha, I; Ferriby, D; Lacour, A; Labalette, P; de Seze, J; Vermersch, P. Optical coherence tomography in clinically isolated syndrome: no evidence of subclinical retinal axonal loss. *Arch. Neurol.* 2009; 66, 1373-1377.

[44] Oberwahrenbrock, T; Schippling, S; Ringelstein, M; Kaufhold, F; Zimmermann, H; Keser, N; Young, KL; Harmel, J; Hartung, HP; Martin, R; Paul, F; Aktas, O; Brandt, AU. Retinal damage in multiple sclerosis disease subtypes measured by high-resolution optical coherence tomography. *Mult. Scler. Int* 2012; 2012, 530305.

[45] Syc, SB; Saidha, S; Newsome, SD; Ratchford, JN; Levy, M; Ford, E; Crainiceanu, CM; Durbin, MK; Oakley, JD; Meyer, SA; Frohman, EM; Calabresi, PA. Optical coherence tomography segmentation reveals ganglion cell layer pathology after optic neuritis. *Brain* 2012; 135, 521-533.

[46] Fjeldstad, C; Bemben, M; Pardo, G. Reduced retinal nerve fiber layer and macular thickness in patients with multiple sclerosis with no history of optic neuritis identified by the use of spectral domain high-definition optical coherence tomography. *J. Clin. Neurosci.* 2011; 18, 1469-1472.

[47] Garcia-Martin, E; Pueyo, V; Almarcegui, C; Martin, J; Ara, JR; Sancho, E; Pablo, LE; Dolz, I; Fernandez, J. Risk factors for progressive axonal degeneration of the retinal nerve fibre layer in multiple sclerosis patients. *Br. J. Ophthalmol.* 2011; 95, 1577-1582.

[48] Serbecic, N; Aboul-Enein, F; Beutelspacher, SC; Vass, C; Kristoferitsch, W; Lassmann, H; Reitner, A; Schmidt-Erfurth, U. High resolution spectral domain optical coherence tomography (SD-OCT) in multiple sclerosis: the first follow up study over two years. *PLoS One* 2011; 6, 19843.

[49] Khanifar, AA; Parlitsis, GJ; Ehrlich, JR; Aaker, GD; D'Amico, DJ; Gauthier, SA; Kiss, S. Retinal nerve fiber layer evaluation in multiple sclerosis with spectral domain optical coherence tomography. *Clin. Ophthalmol.* 2010; 4, 1007-1013.

[50] Sergott, RC; Frohman, E; Glanzman, R; Al-Sabbagh, A. The role of optical coherence tomography in multiple sclerosis: expert panel consensus. *J. Neurol. Sci.* 2007; 263, 3-14.

[51] Talman, LS; Bisker, ER; Sackel, DJ; Long, DA Jr; Galetta, KM; Ratchford, JN; Lile, DJ; Farrell, SK; Loguidice, MJ; Remington, G; Conger, A; Frohman, TC; Jacobs, DA; Markowitz, CE; Cutter, GR;

Ying, GS; Dai, Y; Maguire, MG; Galetta, SL; Frohman, EM; Calabresi, PA; Balcer, LJ. Longitudinal study of vision and retinal nerve fiber layer thickness in multiple sclerosis. *Ann. Neurol.* 2010; 67, 749-760.

[52] Holder, GE; Gale, RP; Acheson, JF; Robson, AG. Electrodiagnostic assessment in optic nerve disease. *Curr. Opin. Neurol.* 2009; 22, 3–10.

[53] Wang, J; Cheng, H; Hu, YS; Tang, RA; Frishman, LJ. The photopic negative response of the flash electroretinogram in multiple sclerosis. *Invest. Ophthalmol Vis. Sci.* 2012; 53, 1315-1323.

[54] Fraser, CL; Holder, GE. Electroretinogram findings in unilateral optic neuritis. *Doc. Ophthalmol.* 2011; 123, 173-178.

[55] Viswanathan, S; Frishman, LJ; Robson, JG; Walters, JW. The photopic negative response of the flash electroretinogram in primary open angle glaucoma. *Invest. Ophthalmol Vis. Sci.* 2001; 42, 514–522.

[56] Berninger, TA; Heider, W. Pattern electroretinograms in optic neuritis during the acute stage and after remission. *Graefes Arch. Clin. Exp. Ophthalmol.* 1990; 228, 410–414.

[57] Papakostopoulos, D; Fotiou, F; Hart, JC; Banerji, NK. The electroretinogram in multiple sclerosis and demyelinating optic neuritis. Electroencephalogr. *Clin. Neurophysiol.* 1989; 74, 1-10.

[58] Stefano, E; Cupini, LM; Rizzo, P; Pierelli, F; Rizzo, PA. Simultaneous recording of pattern electroretinogram (PERG) and visual evoked potential (VEP) in multiple sclerosis. *Acta. Neurol. Belg.* 1991; 91, 20-28.

[59] Walsh, P; Kane, N; Butler, S. The clinical role of evoked potentials. J. *Neurol. Neurosurg. Psychiatry* 2005; 76, 16-22.

[60] Halliday, AM; McDonald, WI; Mushin, J. Delayed visual evoked response in optic neuritis. *Lancet* 1972; 1, 982–985.

[61] Halliday, AM; McDonald, WI; Mushin, J. Visual evoked response in diagnosis of multiple sclerosis. *Br. Med. J.* 1973; 4, 661–664.

[62] Balnytė, R; Ulozienė, I; Rastenytė, D; Vaitkus, A; Malcienė, L; Laučkaitė, K. Diagnostic value of conventional visual evoked potentials applied to patients with multiple sclerosis. *Medicina* 2011; 47, 263-269.

[63] Sisto, D; Trojano, M; Vetrugno, M; Trabucco, T; Iliceto, G; Sborgia, C. Subclinical visual involvement in multiple sclerosis: a study by MRI, VEPs, frequency-doubling perimetry, standard perimetry, and contrast sensitivity. *Invest. Ophthalmol Vis. Sci.* 2005; 46, 1264-1268.

[64] Frederiksen, JL; Petrera, J. Serial visual evoked potentials in 90 untreated patients with acute optic neuritis. *Surv. Ophthalmol.* 1999; 44, S54-62.

[65] Weinstock-Guttman, B; Baier, M; Stockton, R; Weinstock, A; Justinger, T; Munschauer, F; Brownscheidle, C; Williams, J; Fisher, E; Miller, D; Rudick, R. Pattern reversal visual evoked potentials as a measure of visual pathway pathology in multiple sclerosis. *Mult. Scler.* 2003; 9, 529-534.

[66] Alshuaib, WB. Progression of visual evoked potential abnormalities in multiple sclerosis and optic neuritis. *Electromyogr. Clin. Neurophysiol.* 2000; 40, 243-252.

[67] Basar-Eroglu, C; Warecka, K; Schürmann, M; Başar, E. Visual evoked potentials in multiple sclerosis: frequency response shows reduced alpha amplitude. *Int. J. Neurosci.* 1993; 73, 235-258.

[68] Mizota, A; Asaumi, N; Takasoh, M; Adachi-Usami, E. Pattern visual evoked potentials in Japanese patients with multiple sclerosis without history of visual pathway involvement. *Doc. Ophthalmol.* 2007; 115, 105-109.

[69] Frederiksen, JL; Larsson, HB; Olesen, J; Stigsby, B. MRI, VEP, SEP and biothesiometry suggest monosymptomatic acute optic neuritis to be a first manifestation of multiple sclerosis. *Acta. Neurol. Scand.* 1991; 83, 343-350.

[70] Sanders, EA; van Lith GH. Optic neuritis, confirmed by visual evoked response, and the risk for multiple sclerosis: a prospective survey. *J. Neurol. Neurosurg. Psychiatry* 1989; 52, 799-800.

[71] MacFadyen, DJ; Drance, SM; Douglas, GR; Airaksinen, PJ; Mawson, DK; Paty, DW. The retinal nerve fiber layer, neuroretinal rim area, and visual evoked potentials in MS. *Neurology* 1988; 38, 1353-1358.

[72] Miletto, G; Merlihot, JM; Nezri, M; Stanoyevitch, JF; Pernet-Rambaud, C; Houis, B. Comparative study of visual evoked potentials and magnetic resonance imaging in multiple sclerosis. *J. Fr. Ophtalmol.* 1988; 11, 67-73.

[73] Sanders, EA; Reulen, JP; Hogenhuis, LA; van der Velde, EA. Electrophysiological disorders in multiple sclerosis and optic neuritis. *Can. J. Neurol. Sci.* 1985; 12, 308-313.

[74] Neima, D; Regan, D. Pattern visual evoked potentials and spatial vision in retrobulbar neuritis and multiple sclerosis. *Arch. Neurol.* 1984; 41, 198-201.

[75] Asselman, P; Chadwick, DW; Marsden, DC. Visual evoked responses in the diagnosis and management of patients suspected of multiple sclerosis. *Brain* 1975; 98, 261–282.

[76] Samsen, P; Chuenkongkaew, WL; Masayaanon, P; Chirapapaisan, N; Ruangvaravate, N; Loket, S. A comparative study of visual evoked potentials in optic neuritis and optic neuritis with multiple sclerosis. *J Med. Assoc. Thai.* 2007; 90, 313-318.

[77] Grover, LK; Hood, DC; Ghadiali, Q; Grippo, TM; Wenick, AS; Greenstein, VC; Behrens, MM; Odel, JG. A comparison of multifocal and conventional visual evoked potential techniques in patients with optic neuritis/multiple sclerosis. *Doc. Ophthalmol.* 2008; 117, 121-128.

[78] Freedman, MS. 'Time is brain' also in multiple sclerosis. *Mult. Scler.* 2009; 15, 1133-1134.

[79] Kolappan, M; Henderson, AP; Jenkins, TM; Wheeler-Kingshott, CA; Plant, GT; Thompson, AJ; Miller, DH. Assessing structure and function of the afferent visual pathway in multiple sclerosis and associated optic neuritis. *J. Neurol.* 2009; 256, 305-319.

[80] Klistorner, A; Arvind, H; Nguyen, T; Garrick, R; Paine, M; Graham, S; Yiannikas, C. Fellow eye changes in optic neuritis correlate with the risk of multiple sclerosis. *Mult. Scler.* 2009; 15, 928-932.

[81] Feinsod, M; Hoyt, WF. Subclinical optic neuropathy in multiple sclerosis: how early VER components reflect axon loss and conduction defects in optic pathways. *J. Neurol. Neurosurg. Psychiatry* 1975; 38, 1109-1114.

[82] Hidajat, RR; Goode, DH. Normalisation of visual evoked potentials after optic neuritis. *Doc. Ophthalmol.* 2003; 106, 305-309.

[83] Brusa, A; Jones, SJ; Plant, GT. Long-term remyelination after optic neuritis. A 2-year visual evoked potential and psychophysical serial study. *Brain* 2001; 124, 468-479.

[84] Hood, DC; Chen, JY; Yang, EB; Rodarte, C; Wenick, AS; Grippo, TM; Odel, JG; Ritch, R. The role of the multifocal visual evoked potential (mfVEP) latency in understanding optic nerve and retinal diseases. *Trans Am. Ophthalmol. Soc.* 2006; 104, 71-77.

[85] Fraser, C; Klistorner, A; Graham, S; Garrick, R; Billson, F; Grigg, J. Multifocal visual evoked potential latency analysis: predicting progression to multiple sclerosis. *Arch. Neurol.* 2006; 63, 847–850.

[86] Laron, M; Cheng, H; Zhang, B; Schiffman, JS; Tang, RA; Frishman, LJ. Assessing visual pathway function in multiple sclerosis patients with multifocal visual evoked potentials. *Mult. Scler.* 2009; 15, 1431-1441.

[87] Laron, M; Cheng, H; Zhang, B; Schiffman, JS; Tang, RA; Frishman, LJ. Comparison of multifocal visual evoked potential, standard automated

perimetry and optical coherence tomography in assessing visual pathway in multiple sclerosis patients. *Mult. Scler.* 2010; 16, 412-426.

[88] Klistorner, A; Fraser, C; Garrick, R; Graham, S; Arvind, H. Correlation between full-field and multifocal VEPs in optic neuritis. *Doc. Ophthalmol.* 2008; 116, 19–27.

[89] Fortune, B; Demirel, S; Zhang, X; Hood, DC; Johnson, CA. Repeatability of normal multifocal VEP: implications for detecting progression. *J. Glaucoma*. 2006; 15, 131–141.

[90] Lascano, AM; Brodbeck, V; Lalive, PH; Chofflon, M; Seeck, M; Michel, CM. Increasing the diagnostic value of evoked potentials in multiple sclerosis by quantitative topographic analysis of multichannel recordings. *J. Clin. Neurophysiol.* 2009; 26, 316-325.

[91] Hood, DC; Odel, JG; Winn, BJ. The multifocal visual evoked potential. *J. Neuro-Ophthalmol.* 2003; 23, 279–289.

[92] Hood, DC; Odel, JG; Zhang, X. Tracking the recovery of local optic nerve function after optic neuritis: a multifocal VEP study. *Invest. Ophthalmol Vis. Sci.* 2000; 41, 4032–4038.

[93] Hood, DC; Ohri, N; Yang, EB; Rodarte, C; Zhang, X; Fortune, B; Johnson, CA. Determining abnormal latencies of multifocal visual evoked potentials: a monocular analysis. *Doc. Ophthalmol.* 2004; 109, 189–199.

[94] Ruseckaite, R; Maddess, T; Danta, G; Lueck, CJ; James, AC. Sparse multifocal stimuli for the detection of multiple sclerosis. *Ann. Neurol.* 2005; 57, 904–913.

In: Advances in Neuropathy Research
Editors: M. O. Brown and B. A. Budd
ISBN: 978-1-62618-400-8

Chapter 4

The Neuropathy Induced by Organophosphorus Compounds

***Guilherme Luz Emerick*[1,2*] *and Lais Silva Fernandes*[1]**
[1] Departamento de Análises Clínicas, Toxicológicas e Bromatológicas, Faculdade de Ciências Farmacêuticas de Ribeirão Preto-USP, Ribeirão Preto, SP, Brazil
[2] Instituto de Ciências da Saúde-ICS, Universidade Federal do Mato Grosso-MT, Campus Sinop-MT, Sinop, MT, Brasil

Abstract

Organophosphorus-induced delayed neuropathy (OPIDN) was observed in more than 70,000 human cases in the last 75 years. This type of neuropathology has been known since 1930s when the tuberculosis patients were treated with phosphocreosote, which is an undefined mixture of esters derived from phosphoric acid and phenols. In the 1930s, thousands of cases of paralysis occurred in the United States due to the consumption of Jamaica ginger extract contaminated with tri-ortho-cresyl phosphate (TOCP), which is the compound used as inducer in

* Corresponding Author: Guilherme L. Emerick. Present address: Instituto de Ciências da Saúde-ICS, Universidade Federal do Mato Grosso-MT, Avenida Alexandre Ferronato, 1200, 78557-267, Campus Sinop-MT, Sinop, MT, Brazil. E-mail address: glemerick@yahoo.com.br. Phone: +55 66 3531-9796. Fax: +55 66 3531-9796.

experimental studies of delayed neuropathy induced by organophosphates (OPs). OPIDN is characterized by a distal axonopathy, central and peripheral Wallerian-type which develops from 8 to 14 days after intoxication by a neuropathic OP. Among the OPs that cause OPIDN include phosphates, phosphonates and phosphoramidates. TOCP, methamidophos, mipafox, dichlorvos, trichlorfon and leptophos are examples of some compounds that have been reported as inducers of OPIDN. The methamidophos (O,S-dimethyl phosphoramidothioate) has been reported the pesticide as more often causes OPIDN in humans. To develop this neuropathy is required at least 70-80% of inhibition of an esterase known as neuropathy target esterase (NTE). NTE was originally identified by M. K. Johnson as target of neuropathic OPS. NTE is a membrane-bound enzyme with 1327 amino acids belonging to the group of esterases with a molecular weight of 155 kDa and its physiological role is not yet established. The usual symptoms of OPIDN are early cramps, muscle pain, tingling, numbness, weakness and progressive distal burning more present in feet than in hands. Then, a progressive weakness in the legs is evident. More severe cases can lead to paralysis, respiratory problems and death. Several animal species are sensitive to OPIDN. In this list include chickens, cats, dogs, horses, sheep, pigs, monkeys and man. The hens are the animal model for studying this neuropathy, due to their sensitivity and clinical signs were very similar to those seen in humans. Furthermore, the progress of these signals is easily observed and correlated with pathological changes.

Keywords: Organophosphate pesticides, OPIDN, axonopathy

1. Introduction

The widespread use of pesticides in agriculture in order to increase crop production has been generating a series of toxicological problems that are sometimes misdiagnosed [1]. The main trouble pesticides with regard to acute poisoning globally has been [2] organophosphorus (OPs), carbamates and paraquat. Presently, most organochlorines (OCs) [3] insecticides are banned or severely restricted, including in developing countries, thanks to environmentalists who successfully lobbied that their persistence and bioaccumulation implied unacceptable long-term environmental concerns [4]. During the 1960s and 70s, OP and subsequently carbamate insecticides were introduced in a large scale on the world market. These cholinesterase's inhibiting pesticides were assumed to be less environmentally damaging but with higher acute toxicity than the OCs [5].

OPs are derived from phosphoric, phosphonic and phosphinic acids (see Figure 1). The French chemists, Jean Louis Lassaigne and Philip de Clermount are credited with the synthesis of the first OPs in the nineteenth century, while the initial development of OPs as insecticides and chemical warfare agents early in the twentieth century is primarily attributed to the German chemist Gerhard Schrader [6].

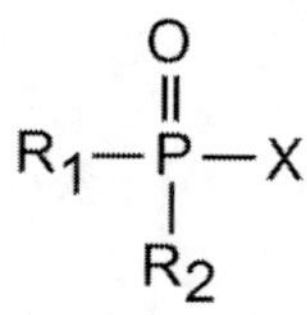

Figure 1. General Structure of OP Insecticides.

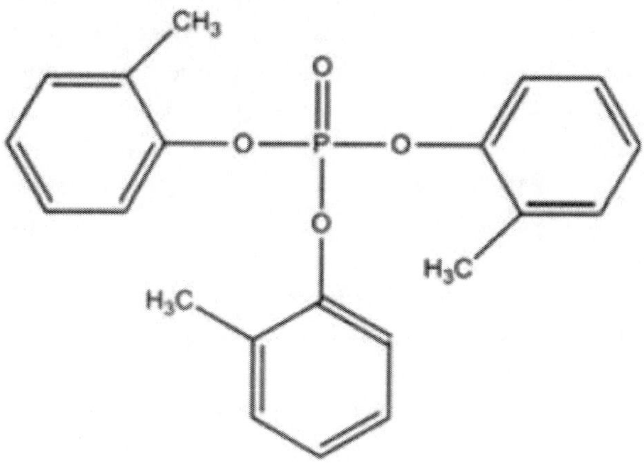

Figure 2. Structure of the Tri-ortho-cresyl phosphate (TOCP).

OP compounds are a group of highly toxic chemicals that have been used widely as pesticides and some were developed as chemical warfare nerve agents such as soman (GD), sarin (GB) and tabun [7]. OPs can be efficiently absorbed by inhalation, ingestion, and skin penetration [8]. OP poisoning resulting from accidental or intentional (suicides attempts) exposure causes approximately one million cases of poisoning each year with several hundred thousand deaths, mostly in developing countries [9, 10].

The use of OP-based pesticides (specifically) is now considered a worldwide health problem since they are the most commonly used and the most often associated with toxicity to humans [11, 12]. In the United States (US) alone, the most recent estimates available (from 2007) indicates that approximately 33 million pounds of OP-based pesticides are applied annually [13]. Internationally, OP-pesticide poisonings are among the most common modes of poisoning-related fatalities (i.e., both intentional and unintentional), a phenomenon that has been attributed to ease of access to OPs and the relatively low level of regulations governing for their use, especially in develop- ing countries. Clinical manifestations of OP intoxication may vary among victims, depending on the specific compound, the amount, route and duration of exposure as well as the age and health status of the person exposed.

Clinically, the symptoms of OP poisoning are divided into four syndromes: acute cholinergic syndrome, intermediate syndrome (IS), chronic organophosphorus-induced neuropsychiatric disorders (COPIND) and organophosphorus-induced delayed neuropathy (OPIDN) [14].

OPIDN has been observed in more than 70,000 human cases in the last 75 years. This neuropathy has been known since the 1930s, when occurred outbreak so-called Ginger-Jake paralysis due to the consumption of Jamaica ginger extract contaminated with tri-ortho-cresyl phosphate (TOCP) (Figure 2) [15].

The mechanism of the acute toxicity of OPs in both target and non-target organisms is primarily attributed to inhibitory actions on various forms of cholinesterase leading to excessive peripheral and central cholinergic activity. However, there is now substantial evidence that this canonical (cholinesterase-based) mechanism cannot alone account for the wide-variety of adverse consequences of OP exposure that have been described, especially those associated with repeated exposures to levels that produce no overt signs of acute toxicity [6]. Therefore, it is necessary to understand the mechanism of toxic action of these compounds to prevent serious and irreversible effects.

2. Chemistry of Organophosphorus Pesticides

Pesticides can be defined as any substance or mixture o substances intended for preventing, destroying, repelling, or mitigating pests. Pests can be insects, rodents, weeds, and a host of other unwanted organisms [16].

The development of the OP compounds as pesticides only occurred in the late 1930s and early 1940s when the German chemist Gerhard Schrader discovery the general chemist structure of anticholinesterases OP pesticides.

This chemist also synthesized the first commercialized OP pesticide, called Bladan, containing TEPP (tetraethyl pyrophosphate) as the active ingredient and one of the most known, parathion, in 1994 (Figure 3) [7].

The general structure of OP insecticides can be observed in the Figure 1, where X is the so-called "leaving group", that is displaced when the OP phosphorylates acetylcholinesterase (AChE) [16] and can be any one of a wide variety of substituted aliphatic, aromatic or heterocyclic groups, linked to phosphorus via a bond of some lability (usually -O- or -S-) [17]. In the Figure 1, R1 and R2 are usually simple alkyl or aryl groups bonded directly to the phosphorus atom (phosphinates), or linked via -O- or -S- (phosphates). In phosphonates, R1 may be bonded directly and R2 via one of the above groups. When the carbon is linked to phosphorus atom through an amino group the name is phosphoramidate. In some OPs that contain a sulfur bound to the phosphorus (P=S), metabolic bioactivation is necessary for their biological activity, because only compounds that have a P=O moiety are effective inhibitors of AChE [18].

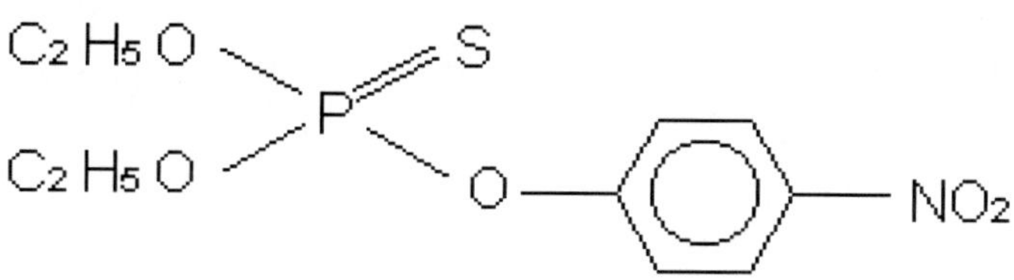

Figure 3. Structure of the Parathion.

The organophosphate insecticides are absorbed through the skin, respiratory and gastrointestinal tract, and often their absorption is favored by solvents present in the formulation. The absorption through the skin is higher at higher temperatures or when there are skin lesions. After absorption, these compounds and their metabolites are rapidly distributed to all tissues [16].

3. Organophosphate Intoxication

The main target of the OP pesticides toxicity is the enzyme acetylcholinesterase (AChE) [19] that is responsible for the rapid and efficient hydrolysis of the neurotransmitter acetylcholine (ACh). Under normal condi- tions, the presynaptic cell synthesizes acetylcholine from choline plus the co-factor acetyl coenzyme A, through the action of the enzyme choline acetyltransferase. Once released, ACh is rapidly removed by acetylchol- inesterase. However, when inhibition of AChE occurs, the efficiency of this enzyme decreases and the extending of acetylcholine effect at synapses is evident [20].

The transmitter acetylcholine is delivered in the synapses of the parasympathetic nervous system and central nervous system (CNS). Cholinergic receptors are divided into muscarinic and nicotinic [21]. Muscarinic cholinergic receptors are widely distributed in various biological systems, where they participate in several vital functions. The activation of these receptors in the peripheral nervous system has actions that include reducing the frequency and strength of cardiac contraction, relaxation of blood vessels and peripheral constriction of the airways (bronchi and bronchioles). In the CNS, muscarinic receptors are involved in the control of extrapyramidal function, vestibular, in cognitive functions such as memory, learning and attention, modulation of stress and wakefulness. The nicotinic cholinergic receptors are ionotropic receptors, and when are activated acquire a conformation to open the channel permeable to Na^+ and K^+. When compared to synaptic transmission induced by muscarinic receptors, it is observed that nicotinic receptors mediate one excitatory fast transmission. These receptors are composed of five protein subunits and are distributed in various tissues, including brain and muscle tissue [20, 22].

The OP components act as pseudo substrates for AChE and form an intermediate compound covalently phosphorylated, which is resistant to aqueous hydrolysis. Thus, there is inhibition of AChE activity, and the esterase does not act on its natural substrate, ACh. When this inhibition occurs, there is an accumulation of ACh in the synapses of the central and peripheral nervous system, leading to an overstimulation of cholinergic receptors that exceeds the normal physiologic limits, known as cholinergic syndrome [7].

The stimulation of muscarinic receptors leads to secretions increased, gastrointestinal cramps, bradycardia, miosis, diarrhea and polyuria. On the other hand, nicotinic receptors when stimulated lead to generalized muscle weakness, fatigue, cramps and fasciculations. The central nervous system stimulation leads to mental confusion, headaches, seizures and coma (See Table 1).

The drug in use to treat acute intoxication by OP is atropine that acts as antagonist at muscarinic cholinergic synapses. However, atropine does not bind to nicotinic receptors and may not relieve the nicotinic effects in acute intoxication caused by OP. In addition, the use pralidoxime in high doses exposure is common, because oximes can reactivate AChE by removing the phosphoryl group attached to the ester group of the enzyme. The diazepam is also used in acute intoxication by OP because it relieves anxiety in milder cases and reduces muscle twitching and antagonizes seizures in severe cases [23].

The accumulation of acetylcholine at synapses also results in rapid and profound excitotoxicity, and leads to dysfunction of cholinergic neurons in the brain due to excessive activation of muscarinic receptors. The overstimulation of muscarinic receptors also upsets the imbalance of glutamatergic and GABAergic activity [8]. The cholinergic neuronal excitotoxic and the dysfunction will lead to a loss of integrity of some cholinergic pathways resulting in a variety of negative effects, which subsequently may cause neuronal cell death and neuronal loss in brain cholinergic regions.

Table 1. Signs and Symptoms of Acute Poisoning by organophosphate compounds

Muscarinic receptor	Nicotinic receptor	Central nervous system
Increased secretions	Fatigue	Mental confusion
Bronchoconstriction	Cramps	Headaches
Bradycardia	Generalized muscle weakness	Convulsions
Polyuria	Fasciculations	Coma
Diahrrea	Tachycardia	Dizziness
Gastrointestinal cramping	transient hypertension	depression of respiratory centers
Miosis	twitching	lethargy
lacrimation	flaccid paralysis	
hypotension		

A second clinical manifestation of OP poisoning is called of Intermediate Syndrome. Intermediate syndrome was first described by Senanayake and Karalliedde in 1987 and appears after cholinergic crisis and before OPIDN. This syndrome was characterized by weakness of the proximal muscles of the limbs, neck flexor muscles, respiratory muscles and cranial motor nerves. It has also been attributed to muscle fiber necrosis after acute cholinergic crisis [24]. It is seen in 20-50% of cases of acute poisoning by OP and has been observed after exposure to a wide range of OP. This syndrome develops in 24 to 96 hours after OP intoxication during recovery of cholinergic manifestations, or in some cases when the patient is already completely recovered from the initial cholinergic crisis [25]. Although the incidence of

Intermediate Syndrome can be high and being considered as a major cause of morbidity and mortality due to OPs, the underlying pathophysiology of this syndrome remains unclear. One hypothesis is that muscle weakness can result from nicotinic receptor desensitization due to prolonged cholinergic stimulation. There is no specific treatment for intermediate syndrome and intervention is exclusively for support [26].

Some OP are able to induce long-term neurological damage, regardless of its effects on the inhibition of AChE. These effects are known as chronic neuropsychiatric disorders induced by OPs (COPIND) and Organophosphorus induced delayed neuropathy (OPIDN). Chronic neuropsychiatric disturbances induced by OPs can be caused by repeated exposure to low doses. The mechanism of this syndrome is the least known of the chronic effects caused by OPs, but is know that this mechanism is independent of esterases inhibition. The most common clinical signs include memory loss, difficulty concentrating and learning, anxiety, depression, and extrapyramidal symptoms. The OPIDN characterized by a distal axonopathy, central and peripheral Wallerian-type, which develops from 8 to 14 days after intoxication by a neuropathic OP [27]. This neuropathy will be discussed in more detail in the next topic of this chapter.

4. Organophosphorus-induced Delayed Neuropathy (OPIDN)

OPIDN was observed in more than 70,000 human cases in the last 75 years. This neuropathology has been known since the 30s when tuberculosis patients who had been treated with phosphocreosote, which is an undefined mixture of esters derived from phosphoric acid and phenols. In the 30s, thousands of cases of paralysis occurred in the United States due to the consumption of Jamaica extract ginger contaminated with tri-ortho-cresyl phosphate (TOCP) [28].

There is a latency period of 1-2 weeks before the onset of lower limb weakness that progresses over the next week to ataxia and paralysis. Characterized by a distal axonopathy Wallerian-type, OPIDN develops after intoxication by a neuropathic OP. Among the OPs that cause OPIDN include phosphates, phosphonates and phosphoramidates. TOCP, methamidophos, mipafox, dichlorvos, trichlorfon and leptofós are examples of some compounds that have been reported as inducers of OPIDN. The methamidophos (O, S-dimethyl fosforamidotioato) has been reported as the pesticide more often causer of OPIDN in humans [29, 30].

Neuropathological studies in OPIDN have shown that the primary lesion is a bilateral degenerative change in the levels of distal axons and their terminals, which primarily affects the longer central and peripheral nerve fibers, leading to the breakdown of neuritic segments and myelin sheath (Figure 4) [31].

The early usual symptoms are cramps, muscle pain, tingling, numbness, weakness and progressive distal burning more feet than hands. Then, a progressive weakness in the legs that evolves beginning to atrophy of the distal muscles is evident. More severe cases can lead to paralysis, respiratory problems and death [32].

Several studies conducted over the past 30 years have shown that the target for the development of OPIDN is an esterase membrane bound, that is present in nervous tissue and other tissues (eg. lymphocytes), called of neuropathy target esterase (NTE) [4]. The prerequisite to generate OPIDN is 70-80% of NTE inhibition and aging (reaction described below) [26, 30, 33].

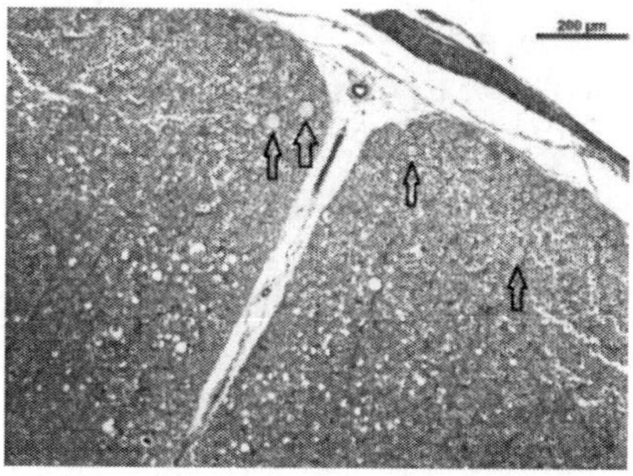

Figure 4. Cross section of spinal cord of hens 21 days after 500 mg/kg of TOCP orally. This section is from lumbar portion and shows the medial pontine-spinal tract with basophilic swollen degenerating axons (arrows). Sections stained with H&E.

Some variables are associated with biochemical mechanism of toxic action of OPs. Some OPs inhibit both AChE and NTE, but do not cause OPIDN, because the concentration required to indice inhibition and aging of more than 70% of the NTE would not be compatible with survival of the individual due to strong cholinergic crisis before the onset of delayed effects. Other OPs like diisopropyl phosphorofluoridate (DFP), mipafox, trichlorfon and dichlorvos, inhibit both AChE and NTE, and are able to induce OPIDN, because the concentration required to induce inhibition and aging of more than 70 % of the NTE are lower or equivalent than the % of AChE inhibition.

The mechanism for the development of OPIDN initially involves the phosphoric dealkylation of NTE. But only the NTE inhibition is not sufficient to result in OPIDN. It is necessary to generate a negative charge in the terminal portion of the enzyme-linked phosphorus grouping, and this occurs as a result of the second reaction, called "aging" [18, 34]. The cleavage of one bond in the R–O–P chain occurs and the loss of R leads to the formation of a charged mono-substituted phosphoric acid residue that is still attached to the protein. This negative compound is extremely stable and resists breakage or hydrolytic reactivation promoted by agents such as oximes [35].

The aging reaction is possible when the radical (R) of the OP is connected to the phosphorus atom via connection for an oxygen atom or a sulfur atom (P–O–R or P–S–R) (Figure 1). The rate of aging enzyme phosphorylated is dependent of radical and is usually slower when they are methyl groups. Certain inhibitors of the NTE such as phosphinates, carbamates, thio-carbamates and sulfonyl fluorides have both radicals "R" directly linked to the phosphorus atom. These compounds also react covalently with the esterase active site, but do not promote the aging reaction therefore esterases inhibited by these compounds do not form a terminal group with a negative charge and are reactivated by endogenous mechanisms [18]. If these NTE non neuropathic inhibitors are given to hens after a neuropathic OP the name is promotion (because they increase the OPIDN signs) but if these inhibitors are given to hens before a neuropathic OP the name is protection (because they protect NTE from neuropathic OPs) [36].

Although NTE inhibition and subsequent aging have been proposed as the initial events of OPIDN, the changes that occur between NTE inhibition and the onset of clinical signs are not completely understood. The proposed mechanism is that NTE inhibition by neuropathics OPs is associated with the decrease in extracellular calcium and the increase of calpain activity (that is activated by the elevated levels of cytoplasmic calcium). This enzyme promotes protein digestion in the terminal portion of axon, which appears to proceed similarly to Wallerian degeneration [37, 38].

Several species are susceptible to OPIDN, such as chickens, cats, dogs, horses, sheep, pigs, monkeys and humans [39]. The adult hen is the animal model for studying this neuropathy (U. S. EPA, 1996), due to their sensitivity and clinical signs were very similar to those seen in humans. Furthermore, the progress of these signals is easily observed and correlated with pathological changes. Still, the chicken provides good reproducibility of results in an animal model relatively small and accessible [40].

4.1. Wallerian-type Degeneration

The concept of axonal degeneration was first reported by Augustus Waller about a hundred years. When the injury occurs by chemical agents in the cell body of the neuron is called of neuronal degeneration. But when the injury occurs in the axon is called axonal degeneration or axonopathy and axon size is an important factor, because the larger myelinated axons are the most affected.

According to the sequence of events that occur in the distal axon, this is called Wallerian-type axonal degeneration and is characterized by activation of calpains due to excessive passage of calcium into the intracellular environment. The activation of these proteases promotes digestion of the terminal portion of the axon, which prevents transmission of nerve impulse to the postsynaptic cell [38].

Initially, there is a period in which the distal axon remains relative structure, transmission and normal activity. The duration of this latent period is proportional to the size of the axon. Following this maintenance period, proteases (calpains) promotes digestion of the axolemma and axoplasm, leaving only the myelin sheath surrounding the swollen axon [41]. Thus, the sequence of OPIDN effects is viewed as a progressive distal axonopathy and starting 8-14 days after exposure to neuropathic OPs.

4.2. Neuropathy Target Esterase

NTE [4] was originally identified by Johnson (1969) [42] as the target of neuropathic OPs, and its inhibition and aging is an early event in the onset of delayed neuropathy in humans and other vertebrates. NTE is a serine esterase linked to the endoplasmic reticulum membrane with 1327 amino acids and molecular weight of 147 kDa. The serine esterases belong to the families of enzymes that have reactive nucleophile at their active site in the catalytic triad together with a histidine and an acidic (aspartate or glutamate) residue.

Probably, NTE has as substrate carboxyl ester compounds and is inhibited in a general mechanism of an esterase by OPs. The active site of the enzyme makes a nucleophilic attack on the acyl-carbon of the ester. The alcoholic hydroxyl is rapidly liberated and a covalent acyl-enzyme compound is produced. This compound is easily hydrolyzed liberating the carboxylic acid and the enzyme. However, when NTE reacts with a neuropathic OP, the rate of hydrolysis of the enzyme-OP is much smaller than the acyl-enzyme, which makes this virtually irreversible inhibition.

The physiological role of NTE is not well defined, but it is known that is related to the homeostasis of phosphatidylcholines. More recently, NTE activity has been evaluated with lysophospholipid substrates more relevant in its physiological function. Lysophospholipids in healthy cells from mammals constitute between 0.5 and 6% of the total weight of lipids. However, in high concentrations, these lipids are linked to atherosclerosis, inflammation and hyperlipidemia [43].

Recent studies proposed that the physiological role of NTE may be related to maintaining of the fluidity of the lipid bilayer of the plasma membrane, and thus the normal activity of NTE is essential for maintaining the neuron membrane and hence the cell permeability [43].

The activity of NTE can be evaluated in the central nervous system and peripheral nerves, as well as in lymphocytes and platelets. The method to evaluate the activity of this enzyme is based on the detection in the visible region (510 nm) of the orange color formed from the reaction of phenol liberated by the breakdown of phenyl valerate (substrate) and potassium ferricyanide [18, 44].

4.3. Genetic Susceptibility

Variations genetically determined in biotransformation enzymes or target molecules can modify the response to an OP. The CYPs are important for the activation and detoxification of OP. Forms of CYPs have been identified and these polymorphisms confer different catalytic activities or expression levels that result in different oxidation rates of OPs [45].

Rainier and co-workers [46] reported the discovery of mutations in genes of the NTE in a consanguineous family and in a not consanguineous family which affected individuals exhibited spastic paraplegia and slowly progressive distal muscular atrophy. In a recent neurogenetics review of two unrelated families, were discovered three distinct mutations in the NTE esterase domain (NEST) for people affected by a form of hereditary spastic paraplegia (SPG39 HSP) which was named NTE-motor neuron disease (NTE-MND) [46].

The motor neuron degeneration in individuals with mutations in NTE supports the role of the biochemical NTE cascade in the pathogenesis of OPIDN and possibly in other neurodegenerative diseases.

4.4. Treatment for OPIDN

Several studies have been conducted in an attempt to relieve the signs and symptoms of OPIDN. Compounds that inhibit NTE, but do not cause the aging reaction have been used to protect NTE from neuropathic OPs. When administered in experimental animals before a neuropathic OP, these compounds exert a protective function, to occupy the active site of NTE [26, 47]. The compound phenylmethylsulfonyl fluoride (PMSF) is the example more common in experimental studies for promotion or protection [48].

Another strategy that has been discussed is to avoid excessive calcium entry into the cell, using calcium channel blockers. The Verapamil, a calcium channel blocker, showed a protective effect in some studies, decreasing phosphorylation of proteins in the brain of chickens receiving TOCP orally [49]. Nimodipine, another calcium channel blocker, decreased phosphorylation of proteins in rats treated with dichlorvos; and moreover, it crosses the blood-brain barrier and acts on the central nervous system [50-52]. Although there are several studies for the treatment of OPIDN, no therapies evaluated experimentally are used in clinical practice for humans.

References

[1] Vasconcellos L. F., Leite A. C., Nascimento O. J. Organophosphate-induced delayed neuropathy: Case report. *Arquivos de neuropsiquiatria.* 2002;60(4):1003-7. Epub 2003/02/04.

[2] Rainier S. A., James W.; Dyck, Peter J.; Eldevik, O. Petter; Wilcock, Sandra; Richardson, Rudy J.; Fink, John K. Motor Neuron Disease Due To Neuropathy Target Esterase Gene Mutation: Clinical Features Of The Index Families. *Muscle & Nerve.* 2010;43(1):19-25.

[3] Agronegócio. MdAPeA. *Agronegócio. Disponível* em: http://www.agriculturagovbr/2009.

[4] Araújo A. J., Lima, J. S., Moreira, J. C., Jacob, S. C., Soares, M. O., Monteiro, M. C. M., Amaral, A. M., Kubota, A., Meyer, A., Cosenza, C. A. N., Neves, C., Markowitz, S. Exposição múltipla a agrotóxicos e efeitos à saúde: Estudo transversal em amostra de 102 trabalhadores rurais. Ciências e Saúde Coletiva, Nova Friburgo, *RJ.* 2007;12:115-30.

[5] Barret D. S., Oehme, F. W., Kruckenberg, S. M. A review of organophosphorus ester induced delayed neurotoxicity. *Veterinary and Human Toxicology.* 1985;27:248-54.

[6] Terry A. V. Jr. Functional consequences of repeated organophosphate exposure: potential non-cholinergic mechanisms. *Pharmacology & therapeutics.* 2012;134(3):355-65. Epub 2012/04/03.

[7] Gallo M. A L. N. *Organic phosphorus pesticides.* In: Hayes WJ LEeSDAP, editor. Handbook of Pesticide Toxicology Academic Press. 1991. p. 917–1123.

[8] Chen Y. Organophosphate-induced brain damage: mechanisms, neuropsychiatric and neurological consequences, and potential therapeutic strategies. *Neurotoxicology.* 2012;33(3):391-400. Epub 2012/04/14.

[9] Pandit V., Seshadri S., Rao S. N., Samarasinghe C., Kumar A., Valsalan R. A case of organophosphate poisoning presenting with seizure and unavailable history of parenteral suicide attempt. *Journal of emergencies, trauma, and shock.* 2011;4(1):132-4. Epub 2011/06/03.

[10] Yurumez Y., Durukan P., Yavuz Y., Ikizceli I., Avsarogullari L., Ozkan S. et al. Acute organophosphate poisoning in university hospital emergency room patients. *Intern Med.* 2007;46(13):965-9. Epub 2007/ 07/03.

[11] Buckley N. A., Roberts D., Eddleston M. Overcoming apathy in research on organophosphate poisoning. *BMJ.* 2004;329(7476):1231-3. Epub 2004/11/20.

[12] Rohlman D. S., Anger W. K., Lein P. J. Correlating neurobehavioral performance with biomarkers of organophosphorous pesticide exposure. *Neurotoxicology.* 2011;32(2):268-76. Epub 2010/12/25.

[13] Agency USEP. *Pesticides industry sales and usage: 2006 and 2007 market estimates.* 2011 (February).

[14] Benson B. J., McIntire, M. Is the intermediate syndrome in organophosphate poisoning the result of insufficient oxime therapy?, *Journal of Toxicology - Clinical Toxicology.* 1992;30:347–9.

[15] Costa L. G. *Toxic effects of pesticides.* In: Klaassen CD, editor. Casarett & Doulls toxicology: the basic science of poisons. New York: McGraw-Hill, 978-0-07-134721-1; 2007. p. 883-930.

[16] Ecobichon D. J. *Toxic effects of pesticides.* In: Klaassen CD, editor. Casarett and Doull's Toxicology:The Basic Science of Poisons. New York: McGraw-Hill, 978-0-07-134721-1; 2001. p. 763-85.

[17] Johnson M. K., Jacobsen, D., Meredith, T. J., Eyer, P., Heath, A. J., Ligtenstein, D. A., Marrs, T. C., Szinicz, L., Vale, J. A., Haines, J. A. Evaluation of antidotes for poisoning by organophosphorus pesticides. *Emergency Medicine.* 2000;12:22-37.

[18] Glynn P. Neural development and neurodegeneration: two faces of neuropathy target esterase. *Progress in neurobiology.* 2000;61(1):61-74. Epub 2000/04/12.

[19] Pope C, Karanth, S., Liu, J. Pharmacology and toxicology of cholinesterase inhibitors: uses and misuses of a common mechanism of action. *Environmental Toxicology and Pharmacology.* 2005;19:433-46.

[20] Glynn P. A mechanism for organophosphate-induced delayed neuropathy. *Toxicology Letters.* 2006;162:94-7.

[21] Read D. J., Li Y., Chao M. V., Cavanagh J. B., Glynn P. Organophosphates induce distal axonal damage, but not brain oedema, by inactivating neuropathy target esterase. *Toxicology and applied pharmacology.* 2010;245(1):108-15. Epub 2010/03/02.

[22] Ana L. M. Ventura P. A. A., Rodrigo C. C. Freitas, Plínio C. Sathler, Natália Loureiro, Helena C. Castro. Sistema colinérgico: revisitando receptores, regulação e a relação com a doença de Alzheimer, esquizofrenia, epilepsia e tabagismo. *Revista de Psiquiatria Clínica.* 2010; 37 (2):66-72.

[23] Glynn P. *Axonal degeneration and neuropathy target esterase.* Arh Hig Rada Toksiko. 2007;58:355-8.

[24] Senanayake N., Karalliedde, L. Neurotoxic effects of organophosphorus insecticides: an intermediate syndrome. *New England Journal of Medicine.* 1987;316:761–3.

[25] Yang C., Deng, J. Intermediate syndrome following organophosphate insecticide poisoning. *Journal of Chinese Medical Association.* 2007; 70:467-72.

[26] Lotti M. *Clinical toxicology of anticholinesterases in humans.* In: Academic Press ----, editor. Handbook of Pesticide Toxicology. San Diego2001.

[27] Tan D., Peng, S., Wu, Y., Wang, Y., Lu, C., Yan, C. Chronic organophosphate (OP)-induced neuropsychiatric disorder is a withdrawal syndrome. *Medical Hypothesis.* 2009;72:405-6.

[28] Zhao X. L., Zhu Z. P., Zhang T. L., Zhang C. L., Yu L. H., Xie K. Q. Tri-ortho-cresyl phosphate (TOCP) decreases the levels of cytoskeletal proteins in hen sciatic nerve. *Toxicology letters.* 2004;152(2):139-47. Epub 2004/08/11.

[29] Mcconell R. T. E., Cuadra R., Torres E., Keifer M., Almendárez J. et al. Organophosphate neuropathy due to methamidophos: biochemical and neurophysiological markers. *Arch Toxicol* 1999;73:296-300.

[30] Johnson M. K. Organophosphorus esters causing delayed neurotoxic effects: mechanism of action and structure activity studies. *Archives of toxicology.* 1975;34(4):259-88. Epub 1975/12/18.

[31] Ehrich M. J. B. *Organophosphorus-induced delayed neuropathy.* In: R K, editor. Handbook of Pesticide Toxicology. San Diego: Academic Press. 2001. p. 987–1012.

[32] Abou-Donia M. B. The cytoskeleton as a target for organophosphorus ester-induced delayed neurotoxicity (OPIDN). Chemico-biological interactions. 1993;87(1-3):383-93. Epub 1993/06/01.

[33] Johnson M. K. G. P. *Neuropathy target esterase.* In: R K, editor. Handbook of Pesticide Toxicology. San Diego: Academic Press 2001. p. 953–65.

[34] Ehrich M., Correll L., Veronesi B. Acetylcholinesterase and neuropathy target esterase inhibitions in neuroblastoma cells to distinguish organophosphorus compounds causing acute and delayed neurotoxicity. Fundamental and applied toxicology : *Official journal of the Society of Toxicology.* 1997;38(1):55-63. Epub 1997/07/01.

[35] Emerick G. L., de Oliveira G. H., dos Santos A. C., Ehrich M. Mechanisms for consideration for intervention in the development of organophosphorus-induced delayed neuropathy. *Chemico-biological interactions.* 2012;199(3):177-84. Epub 2012/07/24.

[36] Lotti M. The Pathogenesis of Organophosphate Polyneuropathy. *Crit Rev Toxicol.* 1992;21(6):465-87.

[37] Inomata M., Hayashi, M., Nakamura, M., Imahori, K., Kawashima, S. Purification and characterization of a calcium-activated neutral protease from rabbit skeletal muscle which requires calcium ions of μM order concentration. *Journal of biochemistry.* 1983;93:291-4.

[38] Anthony D. C., Montine, T. J., Valentine, W. M., Graham, D. G. *Toxic responses of the nervous system.* In: Klaassen CD, editor. Casarett and Doull Toxicology: The Basic Science of Poisons. New York.: McGraw-Hill, 978-0-07-134721-1; 2001. p. 535–59.

[39] Johnson M. K., Glynn P. Neuropathy target esterase (NTE) and organophosphorus-induced delayed polyneuropathy (OPIDP): Recent advances. *Toxicology letters.* 1995;82-83:459-63. Epub 1995/12/01.

[40] Johnson M. K. The primary biochemical lesion leading to the delayed neurotoxic effects of some organophosphorus esters. *Journal of neuro-chemistry.* 1974;23(4):785-9. Epub 1974/10/01.

[41] Jortner B. S., Hancock S. K., Hinckley J., Flory L., Tobias L., Williams L. et al. Neuropathological studies of rats following multiple exposure to

tri-ortho-tolyl phosphate, chlorpyrifos and stress. *Toxicologic pathology.* 2005;33(3):378-85. Epub 2005/04/05.

[42] Johnson M. K. The delayed neurotoxic effect of some organophosphorus compounds. Identification of the phosphorylation site as an esterase. *The Biochemical journal.* 1969;114(4):711-7. Epub 1969/ 10/01.

[43] Greiner A. J., Richardson R. J., Worden R. M., Ofoli R. Y. Influence of lysophospholipid hydrolysis by the catalytic domain of neuropathy target esterase on the fluidity of bilayer lipid membranes. *Biochimica et biophysica acta.* 2010;1798(8):1533-9. Epub 2010/03/30.

[44] Maroni M., Bleecker, M. L. Neuropathy target esterase in human lymphocytes and platelets. *Journal of Applied Toxicology.* 1986;6:1-7.

[45] L. G. C. Pesticide exposure. Differential risk for neurotoxic outcomes due to enzyme polymorphisms. *Clin Occup Environ Med.* 2001;1:511–23.

[46] Rainier S., Bui M., Mark E., Thomas D., Tokarz D., Ming L. et al. Neuropathy target esterase gene mutations cause motor neuron disease. *American journal of human genetics.* 2008;82(3):780-5. Epub 2008/03/ 04.

[47] Lotti M. Promotion of organophosphate induced delayed polyneuropathy by certain esterase inhibitors. *Toxicology.* 2002;181-182:245-8. Epub 2002/12/31.

[48] Johnson M. K., Read D. J. Prophylaxis against and promotion of organophosphate-induced delayed neuropathy by phenyl di-n-pentylphosphinate. *Chemico-biological interactions.* 1993;87(1-3):449-55. Epub 1993/06/01.

[49] Wu Y. J., Li M., Li Y. X., Li W., Dai J. Y., Leng X. F. Verapamil abolished the enhancement of protein phosphorylation of brainstem mitochondria and synaptosomes from the hens dosed with tri-o-cresyl phosphate. *Environ Toxicol Pharmacol.* 2007;24(1):67-71. Epub 2007/ 07/01.

[50] El-Fawal H. A., Correll L., Gay L., Ehrich M. Protease activity in brain, nerve, and muscle of hens given neuropathy-inducing organophosphates and a calcium channel blocker. *Toxicol Appl Pharmacol.* 1990;103(1): 133-42. Epub 1990/03/15.

[51] Emerick E., M., G. L., Jortner, B. S., Oliveira, R. V., de Oliveira, G. H. Biochemical, histopathological and clinical evaluation of delayed effects caused by methamidophos isoforms and tocp in hens: ameliorative effects using control of calcium homeostasis. *Toxicol.* 2012 c;doi: http://dx.doi.org/10.1016/j.tox.2012.08.002.).

[52] Emerick G. L., Peccinini R. G., de Oliveira G. H. Organophosphorus-induced delayed neuropathy: a simple and efficient therapeutic strategy. *Toxicology letters.* 2010;192(2):238-44. Epub 2009/11/17.

In: Advances in Neuropathy Research ISBN: 978-1-62618-400-8
Editors: M. O. Brown and B. A. Budd

Chapter 5

The Role of Neuromuscular Ultrasound in the Diagnostic of Immune-Mediated Neuropathies

Antonios Kerasnoudis [*] ***and Min-Suk Yoon*** [†]
Neurological Department, St. Josef Hospital,
Ruhr University, Bochum, Germany

Abstract

The immune-neuropathies are a diverse group of syndromes. Their diagnosis and classification is based in classical cases, on the distribution pattern of peripheral nerve impairment and the results of nerve conduction studies. In some cases though, the clinical presentation may be subtle, so that extended differential diagnostic workup is needed, including cerebrospinal fluid examination, laboratory tests and in some cases nerve biopsy or magnetic resonance imaging of proximal nerves or roots. Electrophysiological testing remains nowadays fundamental not only for the diagnosis, but also for the follow-up and measurement of response to immune-treatment in these type of neuropathies. New

[*] Antonios Kerasnoudis, MD, Dept. of Neurology, Ruhr University, St. Josef-Hospital, Gudrunstr. 56, 44791 Bochum, Germany, Email: antonis.kerasnoudis@gmail.com, Tel: 0049-17662923235, Fax: 0049-2345093024.

[†] Min-Suk Yoon, MD, Dept. of Neurology, Ruhr University, St. Josef-Hospital, Gudrunstr. 56, 44791 Bochum, Germany, Email: min-suk.yoon@uni-due.de.

challenges arise though, on how best to acquire a static and dynamic imaging of the relevant nerve structures, aiming to provide a complementary and holistic approach to the nerve impairment. As ultrasound technology and resolution continoue to improve, nerve imaging has become more detailed and allows visualisation of nerve fascicles and small structural changes, which will move nerve imaging closer to the type of detail that can be seen in biopsy material. Neuromuscular ultrasound has been able to detect thickened or swollen roots, peripheral nerves or plexus, findings that are consistent with ongoing inflammation, especially in cases of chronic inflammatory demyelinating polyneuropathy (CIDP). Similar findings have been described also in other immunomediated neuropathies such as *Guillain–Barré* syndrome (GBS), multifocal motor neuropathy (MMN) and multifocal acquired demyelinating sensory and motor neuropathy (MADSAM). Although this figures are rewarding to both physicians and patients, it remains unknown, whether ultrasound could be a reliable method for follow up and measurement of response to immune-treatment.

This review provides a timely update on the ultrasound findings of different immune-neuropathies, while possible future possibilities of neuromuscular ultrasound are also discussed.

Introduction

Ultrasound has gained in the last 30 years a key position in the diagnostic approach of the patient with vascular lesions of the central nervous system. The development, however in the last two decades of the high frequency ultrasonography (> 12Mhz), provided the clinician with a tool to study the structures of the peripheral nervous system in detail. The first descriptions of pathological ultrasound findings of the peripheral nerve system have been published in 1985 by Solbiati et al., who reported the pathological changes of recurrent laryngeal nerve lesions in cadavers [1]. In 1988, Fornage and Rifkin described for the first time the ultrasound findings of carpal tunnel syndrome, opening the way for an innovative diagnostic approach to peripheral nerve disease [2]. As ultrasound technology and resolution continuously improve, the scientific interest of neurologists for this method has increased enormously and has been followed by various publications on peripheral nerve disease. Polyneuropathies have an estimated prevalence of about 2% in general population [3, 4]. The immune-mediated neuropathies are a common and diverse group of disorders, resulting in an immune-mediated damage to the peripheral nerves. Barohn et al reported that approximately 13% of consecutive patients with neuropathy seen at their neuromuscular reference centre had an immune-mediated neuropathy [5], while Verghese et al found that 6% of their elderly neuropathy patients had a demyelinating damage of inflammatory etiology [6]. Their diagnosis and classification is based, in typical cases, on the time course, distribution pattern of nerve impairment (predominant involvement of motor/sensory fibers or/and autonomic nerve system), and paraclinical parameters [such as nerve conductions studies and serum antibodies]. On cases, with atypical clinical presentation, an extended diagnostic work-up, including cerebrospinal fluid examination, nerve conductions studies, sural nerve biopsy, laboratory testing, may be needed.

The role of neuromuscular ultrasound in the diagnostic workup of polyneuropathies remains less well defined and parallels the beginning of research on entrapment neuropathies. Only a few studies in the literature have used ultrasound to examine the pathological changes in immune-mediated neuropathies and their correlation with the clinical and electrophysiological findings. The already published case reports/series provide the scientific society with the hope of using neuromuscular ultrasound as a helpful tool in addition to nerve conduction studies and clinical examination for the classification of immune-mediated neuropathies. In this chapter, the basic principles of high frequency ultrasound, along with a literature update on the ultrasound findings of immune-mediated neuropathies are presented.

Basic Principles for Understanding Neuromuscular Ultrasound

Ultrasound is a cyclic sound pressure wave with a frequency higher than the upper limit of human hearing. For the production of ultrasound a phenomenon called piezoelectricity, meaning electricity resulting from pressure is used. The term derives from the Greek piezein (πιέζειν), which means to squeeze or press, and electron (ήλεκτρον), which stands for amber, an ancient source of electric charge. Therefore, a mechanical transducer, containing piezoelectric crystals, is converting electrical signals into mechanical vibrations (ultrasound waves) that are transmitted to human tissues. At the boundaries between two tissues, a part of these ultrasound waves is reflected and received from the transducer again and another part is transmitted in the tissue. Both reflection and transmission depend on the unique structure of every tissue and the transducer frequency. Physically, the higher the frequency of emission, the lower is the penetration of the ultrasound wave deep into the tissue, the more reflection.

In neuromuscular ultrasound, linear array transducers are used for imaging of nerve and muscle tissues. They consist of a row of many single piezoelectric crystals creating a single beam. During this procedure, ultrasound waves, with a frequency of 1-20Mhz, are applied to the skin by a linear transducer. The resolution of the image received, is based on the spatial and tissue resolution. Spatial resolution is defined as the measure of the smallest area identifiable on an image as a discrete separate unit and is further divided into axial and spatial resolution. The *axial resolution* is the minimum distance in the beam direction between two objects (reflectors), which can be identified as separate echoes and is improved when the pulse length decreases or the insonation frequency increases. The *lateral resolution* is defined as the resolution of objects in a plane perpendicular to the axis of an ultrasound beam and can be improved by using a smaller beam width at the focus zone, or by reduction of the frame rate. Image quality in neuromuscular ultrasound has been tremendously improved by the technological development of the ultrasound devices, especially after the discovery of the tissue harmoning imaging technique. According to this technique, images created are derived solely from the higher frequency (called second harmonic sound) produced when the ultrasound pulse passes through tissue within the body. Tissue harmonics, uses various techniques to eliminate the echoes arising from the main transmitted ultrasound beam ("the fundamental frequencies"), from which conventional images are made. Once the fundamental frequencies are eliminated, only the harmonic frequencies are left for image formation. Indeed, the quality of the harmonic image is primarily dependent on the complete elimination of all echoes derived from the transmitted frequencies.

Based on these principles, many peripheral nerves can be easily approached with the use of high frequency ultrasonography. When imaging the nerves, a linear probe with frequency at least > 12 Mhz should be kept perpencidular to the skin surface, avoiding any pressure, that can cause deformity in the structures of interest. The most frequent artifact seen, when receiving ultrasound images of nerves is the anisotropy, meaning the difference, when measured along different axes, in the nerve's physical or mechanical properties. That means the examiner should constantly guide the probe to obtain the best quality of images.

Normal Findings in Peripheral Nerves

Fornage al. published for the first time the normal findings of the most significant peripheral nerves, such as the median, ulnar, sciatic and fibular nerve [2], while Silvestri et al. described in detail the ultrasound structure of these nerves [7]. According to this study, normal peripheral nerves have a tubular form, with alternating hypoechogenic and hyperchogenic zones, corresponding to nerve fibers and perineurium, given the impression of a "honeycomb" pattern when scanning transversely (Figure 1). Their size can be measured accurately in both the longitudinal and transverse scans, and depends mainly on their function (motor, sensory, mixed) [8, 9]. The sonographic differentiation between nerves and tendons is easy, as tendons, in contrast to nerves, appear to have numerous parallel hyperechoic lines, seperated by hypoechoic lines (fibrillar pattern). When scanning longitudinally, the reduced slippage of the nerves during the active or passive bending/extension of a limb, is a further good point in differentiating them from tendons [10].

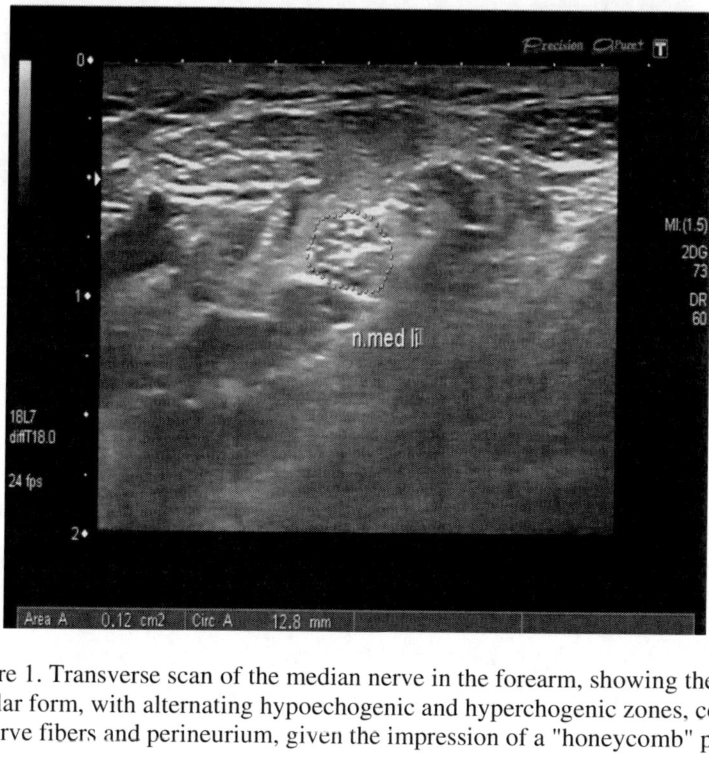

Figure 1. Transverse scan of the median nerve in the forearm, showing the typical tubular form, with alternating hypoechogenic and hyperchogenic zones, corresponding to nerve fibers and perineurium, given the impression of a "honeycomb" pattern.

Quantification of Ultrasound Findings

Cross sectional area (CSA) reference values for peripheral nerves and brachial plexus have been reported in various studies in the literature [11-14]. The difficulty to differentiate a normal from a pathological heterogeneity of cross sectional area changes in peripheral nerves, especially in the case of immune-mediated neuropathies, remains an important limitation of neuromuscular ultrasound in clinical practice. Two novel ultrasound measures, aiming to quantify pathological ultrasound changes of peripheral nerves, have been recently introduced from Padua et al. [15]: 1) the *intranerve cross sectional area variability* (for each nerve), defined as maximal cross sectional area / minimal cross sectional area and 2) the *internerve cross sectional area variability* (for each patient), defined as nerve with maximal intranerve cross sectional area variability / nerve with minimal intranerve cross sectional area variabilty. In addition, our study group introduced **another** ultrasound **measure**: 3) the *side to side difference ratio of the intranerve cross sectional area variability* (SSDIVA) (for each nerve), defined as side with maximal intranerve cross sectional area variability / side with minimal intranerve cross sectional area variability [16] (Figure 2).

Intranerve CSA variability (for each nerve) =	maximal CSA / minimal CSA
Internerve CSA variability (for each subject) =	peripheral nerve with the maximal intranerve CSA variability / peripheral nerve with the minimal intranerve CSA variability
Side to side difference ratio of the intranerve CSA variability (for each nerve) =	side with the maximal intranerve CSA variability / side with the minimal intranerve CSA variability

Figure 2. Equations for calculating the intranerve-, internerve cross sectional area (CSA) variability and side to side difference ratio of the intranerve CSA variability.

Using the intranerve cross sectional area variability the sonographer may differentiate in immune-mediated neuropathies focal (higher values) from diffuse (lower values) nerve enlargement, while the internerve cross sectional area variability may reveal possible distribution patterns of peripheral nerve impairment [15, 16]. On the other hand the *side to side difference ratio of the intranerve cross sectional area variability* may be useful in detecting any lateralization of pathological changes and the intraplexus cross sectional area variability in differentiating focal (higher values) from diffuse (lower values) brachial plexus enlargement.

Ultrasound Findings in Immune-Mediated Neuropathies

1. Chronic Inflammatory Demyelinating Neuropathy (CIDP)

The chronic inflammatory demyelinating polyneuropathy (CIDP) is an acquired immune-mediated inflammatory disorder of the peripheral nerve system. Its estimated prevalence is about 9 per 100,000 adults [17-19]. It is clinically characterized by the progressive or relapsing occurrence of symmetrical weakness in both proximal and distal muscles, impaired sensation, absent or diminished tendon reflexes, an elevated cerebrospinal fluid protein level, demyelinating nerve-conduction studies, and signs of demyelination in nerve-biopsy specimens. The most widely used treatments for CIDP consist of intravenous immunoglobulin, plasma exchange, corticosteroids and immunosuppresion.

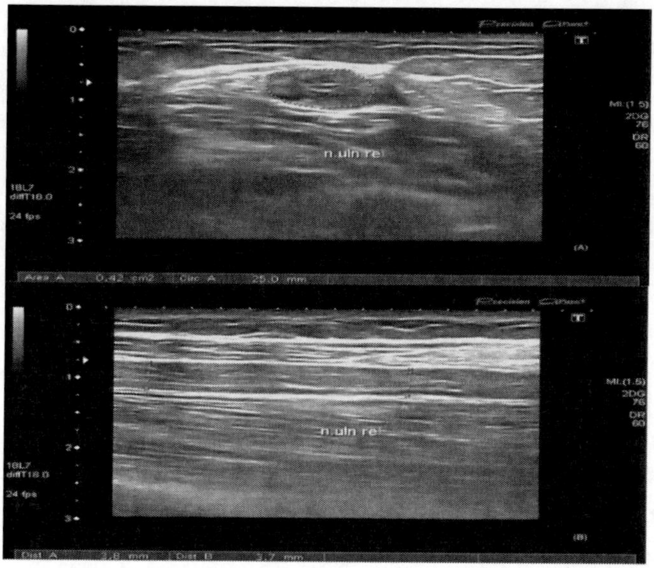

Figure 3. Transverse (A) and longitudinal (B) scan of the ulnar nerve in the forearm in a patient with chronic inflammatory demyelinating polyneuropathy (CIDP) during acute clinical deterioration. Both pictures depict signs of nerve edema, with disturbed fascicular echostructure and cross sectional area enlargement.

The first description of the sonographic findings of CIDP was published from Taniguchi et al. [20]. In this report, the authors documented a brachial plexus hypertrophy on both sides and peripheral nerve hypertrophy at several sites. Similar findings had only been reported in MRI studies until then [21-23]. A possible explanation of this finding could derive from the classical "onion-bulb" histological appearance of the nerves in CIDP, as a result of recurrent episodes of demyelination and remyelination [24]. (Figure 3)

After this initial publication, several years passed until the first systematic ultrasound study of CIDP patients was published. In 2004, Matsuoka et al. reported the ultrasound findings of the cervical roots in 13 patients with CIDP and 35 healthy individuals [25]. The authors demonstrated a hypertrophy of the cervical roots in 9 out of 13 patients with CIDP, a finding that seems to correlate with the elevated levels of protein in CSF. Similar findings have been reported in several case reports in the following years [26-29]. In a further ultrasound study of 36 CIDP patients, diffuse nerve enlargements could be demonstrated [30]. These findings seem to correlate with the disease duration and show an inverse correlation with the NCS findings [31].

Another important aspect in the field of sonography in CIDP, is the possible use of this method for identifying nerve conduction blocks. The localization of the nerve conduction block is often difficult to be made in the nerve conduction studies, especially when dealing with proximal parts of the nerves. By overlooking this typical NCS finding of CIDP, a delay in the diagnosis and therefore treatment can occur. In three CIDP cases in the literature, a correlation between the site of hypertrophy detected with ultrasound and the site of conduction block detected with NCS could be demonstrated [32-34]. Although this seems to be a promising development, it is worth noting that Zaidman et al. failed to confirm these findings in a later study [30]. Systematic studies are therefore required to proof the sensitivity and specificity of this finding.

A novel approach to the quantification of the pathological findings in CIDP was recently published from Padua et al. [15]. Using two new measures, the intranerve- and internerve cross sectional area variability, in a small group of immune-mediated neuropathies, the authors were able to demonstrate that the CIDP shows preferably a diffuse pattern of nerve enlargement (lower values of intranerve cross sectional area variability), when compared to other immune-neuropathies, such as the multifocal motor neuropathy (MMN) (higher values of intranerve cross sectional area variability). These findings have been confirmed in a later study from our ultrasound lab [16].

2. Guillain-Barré-Syndrome (GBS)

GBS is a rare acute inflammatory demylinating polyradiculoneuropathy, occurring at 1–2 cases per 100,000 people annually and is the most common cause of acute non-trauma-related paralysis in the world [36]. Ascending paralysis, weakness beginning in the feet and hands and migrating towards the trunk, is the most typical symptom, and some subtypes cause change in sensation or pain as well as dysfunction of the autonomic nerve system. The diagnosis is primarily based on the typical clinical presentation, while findings of demyelination in nerve conduction studies and CSF signs of cytoalbuminologic dissociation may be absent on the first days of illness.

Very few reports exist about the ultrasound findings of GBS. Zaidman et al. published the findings of 17 patients with GBS, within 4 weeks from the start of neurologic semiology. The median and ulnar nerve found to be a 1.4-fold enlarged compared to controls. There was no correlation, however, between these findings and the nerve conduction monitoring [30]. Similar findings were reported later in a GBS case [37].

3. Multifocal Motor Neuropathy (MMN)

Multifocal motor neuropathy is a progressive, predominantly distal, upper extremity lower motor neuropathy. It is acquired, immune-mediated, and commonly associated with antiganglioside antibodies against GM1. The electrophysiological hallmark is conduction block seen in motor but not sensory nerves. It is ultimately a treatable disease with most patients responding to intravenous immunoglobulin, the treatment of choice.

Beekman et al. published the only existing systematic study on the ultrasound findings of this disease [38]. The authors reported in the majority of the 21 patients examined, a multifocal pattern of nerve enlargement, that was present at sites with and without clinical or electrophysiologic abnormalities. Therefore the sonography and electrophysiologic studies may reveal more abnormalities than expected on purely clinical grounds. A correlation of the site of nerve hypertrophy and site of conduction block in nerve conduction studies has been also reported in an MMN case [39]. Padua et al. reported in a small study higher values of the internerve cross sectional area variability when compared to the subject group, depicting the rather focal distribution pattern of this disease [15]. This pattern could be confirmed in a later prospective study [16]. In addition, the higher values of the side-to-side difference ratio of the cross sectional area variability noted in this disease, may be consistent with the lateralisation of pathological findings expected on clinical means [16].

4. Multifocal Acquired Demyelinating Sensory and Motor Neuropathy (MADSAM)

Multifocal acquired demyelinating sensory and motor (MADSAM) neuropathy is characterized by an asymmetric multifocal pattern of motor and sensory loss, and conduction block and other features of demyelination in nerve conduction studies. This type of neuropathy, also referred to as the Lewis-Sumner syndrome, can be distinguished from multifocal motor neuropathy (MMN) by different clinical presentations and laboratory findings. On the other side, the major features that separate MADSAM neuropathy from typical CIDP are the electrophysiological findings and favorable responses to plasma exchange, prednisone, or IVIg.

There are only small case series in the literature on the ultrasound findings of MADSAM neuropathy. Scheidl et al. reported two MADSAM cases [40]. In the first case, multiple focal nerve enlargements could be detected by ultrasound at sites of previous conduction blocks, well after complete clinical and electrophysiological resolution. In the second case, existing proximal conduction blocks could be localized by ultrasound. This finding, highlighted the importance of nerve ultrasound in identifying conduction blocks and demonstrated that ultrasonographic morphological changes may outlast functional recovery in demyelinating neuropathies such as MADSAM.

In a similar MADSAM case with disease course over 2 years, a hypertrophy of the median and ulnar nerve in the forearm on both sides, but no pathological findings in the lower extremities could be detected. The site where the hypertrophy was detected in ultrasound correlated with the site of conduction block in the nerve conduction studies [41].

The above findings highlight that the pathological changes in MADSAM may be more widespread than expected on clinical and electrophysiologic means. These ultrasound observations could help to understand the underlying mechanisms of nerve damage and facilitate the development of more effective treatments.

5. Vasculitic Neuropathy

Vasculitis, primary or secondary, is the cause of many acquired neuropathies. Peripheral nerves are at high risk of ischemic lesions in vasculitis, which predominantly affect small arteries of the size of the vasa nervorum present in the epineurium of nerve trunks. Peripheral nerve involvement is a common complication of systemic vasculitis, occurring in up to 30% of cases in different series [42]. On classical cases the manifestation can be rapid onset of painful sensory and motor loss affecting the lower extremities or a mononeuritis multiplex.

Nodera et al. observed for the first time a reduction of the echogenicity and diffuse thickening of the ulnar, tibial nerve and the cervical roots in vasculitic neuropathies [43]. In this study ultrasonographic evaluation was repeated after 2 weeks of therapy with corticosteroids and showed significant reduction in the size of the ulnar nerve. No comment was made on the sonographic appearence of the other nerves previously imaged. Ito et al., where able to demonstrate in a ultrasound study of the tibial nerve in patients with vasculitic neuropathy, a statistically significant enlargement of the nerve, when compared to the control group [44]. The authors postulated that this could be the result of vasculitic granulomatous lesions in the epineurium resulting in the development of edema.

Future Directions of Neuromuscular Ultrasound

The growth in the ultrasound technology has provided the clinician with a valuable, secure and non-invasive tool for assessing diseases of the peripheral nerve system. The wide variety of distribution patterns observed in the immune-mediated neuropathies has made it difficult until today to define sufficient criteria for their diagnosis.

On the other hand the difficulty to differentiate a normal from a pathological heterogeneity of cross sectional area alterations in this type of neuropathies, remains an important limitation of neuromuscular ultrasound in clinical praxis. The recently introduced ultrasound measurements, may provide the sonographer with a tool to quantify pathological changes in peripheral nerve disease [15, 16]. The diagnostic value of the proposed measures needs to be tested in different centers in prospective studies.

References

[1] Solbiati L, De Pra L, Ierace T et al. High-resolution sonography of the recurrent laryngeal nerve: anatomic and pathologic considerations. *AJR Am J Roentgenol.* 1985 Nov;145(5):989-93.

[2] Fornage BD, Rifkin MD. Ultrasound examination of the hand and foot. *Radiol Clin North Am.* 1988 Jan;26(1):109-29

[3] Hughes RAC. Peripheral neuropathy. *BMJ* 2002;324:466–469.

[4] England JD, Asbury AK. Peripheral neuropathy. *Lancet* 2004; 363: 2151–2161.

[5] Barohn RJ. Approach to peripheral neuropathy and neuronopathy. *Semin Neurol.* 1998;18(1):7-18.

[6] Verghese J, Bieri PL, Gellido C, Schaumburg HH, Herskovitz S. Peripheral neuropathy in young-old and old-old patients. *Muscle Nerve.* 2001 Nov;24(11):1476-81.

[7] Silvestri E, Martinoli C, Derchi LE, Bertolotto M. Echotexture of peripheral nerves: correlation between US and histologic findings and criteria to differentiate tendons. *Radiology.* 1995 Oct;197(1):291-6

[8] Walker FO. Principles of Ultrasound. In: Walker FO and Cartwight MS. *Neuromuscular Ultrasound*, Elsevier Saunders, Philadelphia, 1. ed. 2011, p 1-23.

[9] Heinemeyer O, Reimers CD. Ultrasound of radial, ulnar, median, and sciatic nerves in healthy subjects and patients with hereditary motor and sensory neuropathies. *Ultrasound Med Biol.* 1999 Mar;25(3):481-5.

[10] Kamolz LP, Schrögendorfer KF, Rab M et al. The precision of ultrasound imaging and its relevance for carpal tunnel syndrome. *Surg Radiol Anat.* 2001;23(2):117-21.

[11] Cartwright MS, Passmore LV, Yoon JS, Brown ME, Caress JB, Walker FO. Cross-sectional area reference values for nerve ultrasonography. *Muscle Nerve.* 2008 May;37(5):566-71

[12] Cartwright MS, Shin HW, Passmore LV, Walker FO. Ultrasonographic reference values for assessing the normal median nerve in adults. *J Neuroimaging.* 2009 Jan;19(1):47-51

[13] Cartwright MS, Shin HW, Passmore LV, Walker FO. Ultrasonographic findings of the normal ulnar nerve in adults. *Arch Phys Med Rehabil.* 2007 Mar;88(3):394-6

[14] Yalcin E, Onder B, Akyuz M. Ulnar nerve measurements in healthy individuals to obtain reference values. *Rheumatol Int.* 2012 Sep 5. [Epub ahead of print]

[15] Padua L, Martinoli C, Pazzaglia C, Lucchetta M, Granata G, Erra C, Briani C. Intra- and internerve cross-sectional area variability: new ultrasound measures. *Muscle Nerve*. 2012 May;45(5):730-3

[16] Kerasnoudis A. Klasing A., Behrendt V., Gold R., M.-S. Yoon. Letter to the editor '' Intra- and internerve cross-sectional area variability: new ultrasound measures''. Muscle and Nerve. *Muscle Nerve*. 2012 Aug 10. doi: 10.1002/mus.23520

[17] Chiò A., Cocito D., Bottacchi E., Buffa C., Leone M., Plano F. et al. (2007) Idiopathic chronic inflammatory demyelinating polyneuropathy: an epidemiological study in Italy. *J Neurol Neurosurg Psychiatry* 78: 1349–1353

[18] Iijima M., Koike H., Hattori N., Tamakoshi A., Katsuno M., Tanaka F. et al. (2008) Prevalence and incidence rates of chronic inflammatory demyelinating polyneuropathy in the Japanese population. *J Neurol Neurosurg Psychiatry* 79: 1040–1043

[19] Hughes R. Chronic Inflammatory Demyelinating Polyradiculo neuropathy. *J Clin Immunol.* 2010 Apr 15.

[20] Taniguchi N., Itoh K., Wang Y. et al: Sonographic detection of diffuse peripheral nerve hypertrophy in chronic inflammatory demyelinating polyneuropathy. *J Clin Ultrasound* 2000; 28:488-491.

[21] Duggins A.J., McLeod J.G., Pollard J.D. et al: Spinal root and plexus hypertrophy in chronic inflammatory demyelinating polyneuropathy. *Brain* 1999; 122:1383-1390.

[22] Mizuno K., Nagamatsu M., Hattori N. et al: Chronic inflammatory demyelinating polyradiculoneuropathy with diffuse and massive peripheral nerve hypertrophy: distinctive clinical and magnetic resonance imaging features. *Muscle Nerve* 1998; 21:805.

[23] Naganuma M., Doi S., Shima K. et al: *Chronic inflammatory demyelinating polyradiculoneuropathy associated with multifocal nerve hypertrophy: report of a case with MRI study*. Rinsho Shinkeigaku 1991; 31:1186.

[24] Matsuda M., Ikeda S., Sakurai S. et al: Hypertrophic neuritis due to chronic inflammatory demyelinating polyradiculoneuropathy (CIDP): a postmortem pathological study. *Muscle Nerve* 1996; 19:163-169

[25] Matsuoka N., Kohriyama T., Ochi K. et al: Detection of cervical nerve root hypertrophy by ultrasonography in chronic inflammatory demyelinating polyradiculoneuropathy. *J Neurol Sci* 2004; 219:15-21

[26] Imamura K, Tajiri Y, Kowa H, Nakashima K. Peripheral nerve hypertrophy in chronic inflammatory demyelinating

polyradiculoneuropathy detected by ultrasonography. *Intern Med.* 2009;48(7):581-2. Epub 2009 Apr 1.

[27] Jang JH, Yoo JK, Kim BJ. Ultrasonographic findings in chronic inflammatory demyelinating polyneuropathy. *Am J Phys Med Rehabil.* 2012 Jun;91(6):546-7

[28] Kerasnoudis A. Ultrasound findings in a case of chronic inflammatory demyelinating polyneuropathy. *Muscle and Nerve*, 2012 Aug 10. doi: 10.1002/mus.23624

[29] Kerasnoudis A. Ultrasonographic findings in chronic inflammatory demyelinating polyneuropathy. *Am J Phys Med Rehabil.* PHM201343 (accepted, ahead of print).

[30] Zaidman C.M., Al-Lozi M., Pestronk A.: Peripheral nerve size in normal and patients with polyneuropathy: an ultrasound study. *Muscle Nerve* 2009; 40:960-966.

[31] Kerasnoudis A. Gold R., Klasing A., Behrendt V., M.-S. Yoon. *Correlation of morphological changes in CIDP with the disease duration.* 85. Kongress der Deutschen Geselschaft für Neurologie (DGN), 27.09.12, Hamburg Germany. P555

[32] Smith E.C., Hobson-Webb L.D., Massey E.: Nerve ultrasound in motor conduction block: pre- and posttreatment findings. *Muscle Nerve* 2008; 38:1369.

[33] Granata G., Pazzaglia C., Calandro P. et al: Ultrasound visualization of nerve morphological alteration at the site of conduction block. *Muscle Nerve* 2009; 40:1068-1070.

[34] Imamura K., Tajiri Y., Kowa H. et al: Peripheral nerve hypertrophy in chronic inflammatory demyelinating polyradiculoneuropathy detected by ultrasonography.*Intern Med* 2009; 48:581-582.

[35] McGrogan A, Madle GC, Seaman HE, de Vries CS. The epidemiology of Guillain-Barré syndrome worldwide. A systematic literature review. *Neuroepidemiology.* 2009;32(2):150-63

[36] Almeida V, Mariotti P, Veltri S, Erra C, Padua L. Nerve ultrasound follow-up in a child with Guillain-Barré syndrome. *Muscle Nerve.* 2012 Aug;46(2):270-5

[37] Beekman R., van den Berg L.H., Franssen H., et al: Ultrasonography shows extensive nerve enlargements in multifocal motor neuropathy. *Neurology* 2005; 65:305-307.

[38] Kerasnoudis A. Correlation of ultrasound and electrophysiological findings in a case of multifocal motor neuropathy. *J Neuroimaging.* 2012 Dec 17. doi: 10.1111/j.1552-6569.2012.00757.x. [Epub ahead of print]

[39] Scheidl E, Böhm J, Simó M, Rózsa C, Bereznai B, Kovács T, Arányi Z. Ultrasonography of MADSAM neuropathy: focal nerve enlargements at sites of existing and resolved conduction blocks. *Neuromuscul Disord.* 2012 Jul;22(7):627-31

[40] Kerasnoudis A. Ultrasonography of MADSAM neuropathy: focal nerve enlargements at sites of existing and resolved conduction blocks. *Neuromuscul Disord.* 2012 Nov;22(11):1032. doi: 10.1016/j.nmd.2012.06.009.

[41] Collins MP, Kissel JT. Vasculitis of the peripheral nervous system. In: Noseworthy JH, ed. *Neurological Therapeutics: Principles and Practice.* London: Martin Dunitz; 2003:2078 –2110

[42] Nodera H., Sato K., Terasawa Y. et al: High-resolution sonography detects inflammatory changes in vasculitic neuropathy. *Muscle Nerve* 2006; 34:380-381.

[43] Ito T., Kijima M., Watanabe T., et al: Ultrasonography of the tibial nerve in vasculitic neuropathy. *Muscle Nerve* 2007; 35:379-382.

Index

A

D

E

F

G

H

I

J

K

L

M

N

O

P

Q

R

S

T